Dirac Operators and Spectral Geometry

The Dirac operator has many useful applications in theoretical physics and mathematics. This book provides a clear, concise and self-contained introduction to the global theory of the Dirac operator and to the analysis of spectral asymptotics with local or non-local boundary conditions.

The theory is introduced at a level suitable for graduate students. Numerous examples are then given to illustrate the peculiar properties of the Dirac operator, and the role of boundary conditions in heat-kernel asymptotics and quantum field theory. Topics covered include the introduction of spin-structures in Riemannian and Lorentzian manifolds; applications of index theory; heat-kernel asymptotics for operators of Laplace type; quark boundary conditions; one-loop quantum cosmology; conformally covariant operators; and the role of the Dirac operator in some recent investigations of four-manifolds.

This volume provides graduate students with a rigorous introduction and researchers with a valuable reference to the Dirac operator and its applications in theoretical physics.

GIAMPIERO ESPOSITO has written three previous books, including the first monograph on Euclidean quantum gravity ever published. His research interests are primarily associated with several problems in classical and quantum gravity. His recent collaborations have led to the discovery of a new class of invariants in the one-loop semiclassical theory, with a corresponding family of universal functions. He received his PhD from the University of Cambridge and held post-doctoral positions in Naples (with the INFN) and Trieste (at the ICTP) before obtaining a permanent position at the INFN in Naples.

Dirac Operators and Spectral Geometry

[illegible]

[illegible]

[illegible]

CAMBRIDGE LECTURE NOTES IN PHYSICS 12
General Editors: P. Goddard, J. Yeomans

1. Clarke: The Analysis of Space-Time Singularities
3. Sciama: Modern Cosmology and the Dark Matter Problem
4. Veltman: Diagrammatica – The Path to Feynman Rules
5. Cardy: Scaling and Renormalization in Statistical Physics
6. Heusler: Black Hole Uniqueness Theorems
7. Coles and Ellis: Is the Universe Open or Closed?
8. Razumov and Saveliev: Lie Algebras, Geometry, and Toda-type Systems
9. Forshaw and Ross: Quantum Chromodynamics and the Pomeron
10. Jensen: Self-organised Criticality
11. Vanderzande: Lattice Models of Polymers
12. Esposito: Dirac Operators and Spectral Geometry

Dirac Operators and Spectral Geometry

GIAMPIERO ESPOSITO

CAMBRIDGE UNIVERSITY PRESS
Cambridge, New York, Melbourne, Madrid, Cape Town, Singapore, São Paulo

Cambridge University Press
The Edinburgh Building, Cambridge CB2 2RU, UK

Published in the United States of America by Cambridge University Press, New York

www.cambridge.org
Information on this title: www.cambridge.org/9780521648622

First published 1998

A catalogue record for this publication is available from the British Library

ISBN-13 978-0-521-64862-2 paperback
ISBN-10 0-521-64862-9 paperback

Transferred to digital printing 2006

This book is dedicated to all those who look for harmony and beauty in the laws of nature

Contents

Preface

The idea of writing some detailed notes on the Dirac operator, and possibly a monograph, came to my mind after I was invited to give a series of graduate lectures at SISSA by Mauro Carfora, in April 1994. Indeed, I had been a graduate student of Cambridge University at St. John's College from 1987 to 1991, sharing my time between the Department of Applied Mathematics and Theoretical Physics and St. John's College, working under the supervision of a distinguished scientist who had, himself, taken great inspiration from Dirac's work. Further to this, the main original results of my Ph.D. thesis were obtained within the framework of local or non-local boundary conditions for the Dirac operator, with application to heat-kernel asymptotics and one-loop quantum cosmology. The problem of boundary conditions in the formulation of physical laws, the interplay of geometric and analytic concepts, and the application of spinorial techniques have always been at the heart of my research since those early days.

Hence I felt it was time to revise substantially my original SISSA lecture notes (early versions appeared as reports GR-QC 9507046 and HEP-TH 9704016) and turn them into a book in the form of a lecture notes monograph for Ph.D. students and research workers. The expert reader will recognize the deep influence of my main sources. Indeed, it is the author's personal opinion that the best way to understand current developments in mathematics and physics is to follow the presentation given, in the original papers, by the leading authors in the various fields. Their clarity and their deep mathematical and physical insight have really no match, and help a lot in the course of forming new generations of scientists, whose task will be the one of elaborating yet new models and new theories, relying on what was achieved before. My first aim has been therefore to perform a careful selection of topics that can be covered in a few weeks of hard work (assuming that the class has already attended introductory courses in fibre-bundle theory and classical groups): What is a spin-structure in Riemannian or Lorentzian geometry? How to build a global theory of the Dirac operator? Why is it so important in the theory of elliptic operators on Riemannian manifolds? What is the meaning of index theory for mathematics and physics? How to extend index theorems to manifolds with boundary? What is the role played by spectral geometry in the theory of quantized fields? The first five chapters are devoted to these issues, jointly with the analysis of non-local boundary conditions for massless spin-1/2 fields, and with a detailed description of operators of Laplace type on manifolds with boundary.

In the concluding chapter, I present a series of non-trivial applications of the general theory of the Dirac operator and of spectral asymptotics: quark boundary conditions, one-loop quantum cosmology, the formalism of conformally covariant operators, and heat-kernel asymptotics in Euclidean quantum gravity. These topics are still the object of intensive investigation by mathematicians and physicists: the development of K-theory and index theory on manifolds with boundary, and the analysis of heat-kernel asymptotics with mixed boundary conditions, are not yet completely understood. By contrast, decades of new fascinating research are in sight. The book ends with a review of the role of the Dirac operator in the derivation of the Seiberg–Witten equations, which have been found to be crucial in the analysis of geometric and topological properties of four-manifolds. I would be satisfied if, further to the key ideas and techniques, I have been able to transmit a sense of excitement and an increasing intellectual curiosity to the general reader. Indeed, as Dirac predicted in his beautiful book on the principles of quantum mechanics, further developments in theoretical physics are likely to involve new and deep ideas in mathematics. The techniques and results described in the present monograph should provide relevant examples of this property.

Giampiero Esposito

Naples

Acknowledgments

I am very grateful to Mauro Carfora for inviting me to SISSA in April 1994. I am indebted to Ivan Avramidi, Peter D'Eath, Alexander Kamenshchik, Klaus Kirsten, Igor Mishakov, Hugo Morales-Técotl and Luis Octavio Pimentel for scientific collaboration, and to Luis Alvarez-Gaumé, Andrei Barvinsky, Stuart Dowker, Mike Eastwood, Peter Gilkey, Hugh Luckock, Ian Moss, Hugh Osborn, Stephen Poletti, Richard Rennie and Dmitri Vassilevich for correspondence. I should also say that the interaction with my former students Gabriele Gionti, Antonella Liccardo and Giuseppe Pollifrone was a source of motivation and inspiration for my research. Last, but not least, it is a pleasure to thank Adam Black and Alison Woollatt of Cambridge University Press, who helped me a lot with the scientific and technical aspects of my manuscript, respectively. My warm thanks are also due to Beverley Lawrence and Jo Clegg for assistance in the copy-editing stage.

1
The Dirac operator

This introductory chapter begins with a review of Dirac's original derivation of the relativistic wave equation for the electron. The emphasis in then put on spin-structures in Lorentzian and Riemannian manifolds, and on the global theory of the Dirac operator, to prepare the ground for the topics studied in the following chapters.

1.1 The Dirac equation

In the development of theoretical physics, the Dirac operator finds its natural place in the attempt to obtain a relativistic wave equation for the electron. Although we shall be mainly concerned with elliptic operators in the rest of our book, we thus devote this section to Dirac's original derivation (Dirac 1928, 1958). Indeed, it is well known that the relativistic Hamiltonian of a particle of mass m is

$$H = c\sqrt{m^2c^2 + p_1^2 + p_2^2 + p_3^2}, \tag{1.1.1}$$

and this leads to the wave equation

$$\left[p_0 - \sqrt{m^2c^2 + p_1^2 + p_2^2 + p_3^2}\right]\psi = 0, \tag{1.1.2}$$

where the p_μ should be regarded as operators:

$$p_\mu \equiv i\hbar\frac{\partial}{\partial x^\mu}. \tag{1.1.3}$$

Thus, multiplication of Eq. (1.1.2) on the left by the operator

$$p_0 + \sqrt{m^2c^2 + p_1^2 + p_2^2 + p_3^2}$$

leads to the equation

$$\left[p_0^2 - m^2c^2 - p_1^2 - p_2^2 - p_3^2\right]\psi = 0, \tag{1.1.4}$$

which is a more appropriate starting point for a relativistic theory. Of course, Eqs. (1.1.2) and (1.1.4) are not completely equivalent: every solution of Eq. (1.1.2) is also, by construction, a solution of Eq. (1.1.4), whereas the converse does not hold. Only the solutions of Eq. (1.1.4) corresponding to positive values of p_0 are also solutions of Eq. (1.1.2).

However, Eq. (1.1.4), being quadratic in p_0, is not of the form desirable in quantum mechanics, where, since the derivation of Schrödinger's equation

$$i\hbar \frac{\partial \psi}{\partial t} = H\psi, \tag{1.1.5}$$

one is familiar with the need to obtain wave equations which are linear in p_0. To combine relativistic invariance with linearity in p_0, and to obtain equivalence with Eq. (1.1.4), one looks for a wave equation which is rational and linear in p_0, p_1, p_2, p_3:

$$\Big[p_0 - \alpha_1 p_1 - \alpha_2 p_2 - \alpha_3 p_3 - \beta\Big]\psi = 0, \tag{1.1.6}$$

where α and β are independent of p. Since we are studying the (idealized) case when the electron moves in the absence of electromagnetic field, all space-time points are equivalent, and hence the operator in square brackets in Eq. (1.1.6) is independent of x. This implies in turn that α and β are independent of x, and commute with the p and x operators. At a deeper level, α and β make it possible to obtain a relativistic description of the spin of the electron, i.e. an angular momentum which does not result from the translational motion of the electron.

At this stage, one can multiply Eq. (1.1.6) on the left by the operator

$$p_0 + \alpha_1 p_1 + \alpha_2 p_2 + \alpha_3 p_3 + \beta.$$

This leads to the equation

$$\Big[p_0^2 - \sum_{i=1}^{3} \alpha_i^2 p_i^2 - \sum_{i \neq j} (\alpha_i \, \alpha_j + \alpha_j \, \alpha_i) p_i p_j$$
$$- \sum_{i=1}^{3} (\alpha_i \, \beta + \beta \, \alpha_i) p_i - \beta^2\Big]\psi = 0. \tag{1.1.7}$$

Equations (1.1.4) and (1.1.7) agree if α_i and β satisfy the conditions

$$\alpha_i^2 = 1 \;\; \forall i = 1, 2, 3, \tag{1.1.8}$$

$$\alpha_i \, \alpha_j + \alpha_j \, \alpha_i = 0 \text{ if } i \neq j, \tag{1.1.9}$$

$$\beta^2 = m^2 c^2, \tag{1.1.10}$$

$$\alpha_i \, \beta + \beta \, \alpha_i = 0 \ \forall i = 1, 2, 3. \tag{1.1.11}$$

Thus, on setting

$$\beta = \alpha_0 mc, \tag{1.1.12}$$

it is possible to re-express the properties (1.1.8)–(1.1.11) by the single equation

$$\alpha_a \, \alpha_b + \alpha_b \, \alpha_a = 2\delta_{ab} \ \forall a, b = 0, 1, 2, 3. \tag{1.1.13}$$

So far, we have found that, if Eq. (1.1.13) holds, the wave equation (1.1.6) is equivalent to Eq. (1.1.4). Thus, one can *assume* that Eq. (1.1.6) is the correct relativistic wave equation for the motion of an electron in the absence of a field. However, Eq. (1.1.6) is not entirely equivalent to Eq. (1.1.2), but, as Dirac first pointed out, it allows for solutions corresponding to negative as well as positive values of p_0. The former are relevant for the theory of positrons, and will not be discussed in our monograph, which focuses instead on physical and mathematical problems in the theory of elliptic operators.

To obtain a representation of four anti-commuting α, as in Eq. (1.1.13), one has to consider 4×4 matrices. Following Dirac, it is convenient to express the α in terms of generalized Pauli matrices (see below), here denoted by $\sigma_1, \sigma_2, \sigma_3$, and of a second set of anti-commuting matrices, ρ_1, ρ_2, ρ_3 say. Explicitly, one may take

$$\alpha_1 \equiv \rho_1 \, \sigma_1, \tag{1.1.14}$$

$$\alpha_2 \equiv \rho_1 \, \sigma_2, \tag{1.1.15}$$

$$\alpha_3 \equiv \rho_1 \, \sigma_3, \tag{1.1.16}$$

$$\alpha_0 \equiv \rho_3, \tag{1.1.17}$$

where (Dirac 1958)

$$\sigma_1 \equiv \begin{pmatrix} 0 & 1 & 0 & 0 \\ 1 & 0 & 0 & 0 \\ 0 & 0 & 0 & 1 \\ 0 & 0 & 1 & 0 \end{pmatrix}, \tag{1.1.18}$$

$$\sigma_2 \equiv \begin{pmatrix} 0 & -i & 0 & 0 \\ i & 0 & 0 & 0 \\ 0 & 0 & 0 & -i \\ 0 & 0 & i & 0 \end{pmatrix}, \tag{1.1.19}$$

$$\sigma_3 \equiv \begin{pmatrix} 1 & 0 & 0 & 0 \\ 0 & -1 & 0 & 0 \\ 0 & 0 & 1 & 0 \\ 0 & 0 & 0 & -1 \end{pmatrix}, \tag{1.1.20}$$

$$\rho_1 \equiv \begin{pmatrix} 0 & 0 & 1 & 0 \\ 0 & 0 & 0 & 1 \\ 1 & 0 & 0 & 0 \\ 0 & 1 & 0 & 0 \end{pmatrix}, \tag{1.1.21}$$

$$\rho_2 \equiv \begin{pmatrix} 0 & 0 & -i & 0 \\ 0 & 0 & 0 & -i \\ i & 0 & 0 & 0 \\ 0 & i & 0 & 0 \end{pmatrix}, \tag{1.1.22}$$

$$\rho_3 \equiv \begin{pmatrix} 1 & 0 & 0 & 0 \\ 0 & 1 & 0 & 0 \\ 0 & 0 & -1 & 0 \\ 0 & 0 & 0 & -1 \end{pmatrix}. \tag{1.1.23}$$

Note that both the ρ and the σ are Hermitian, and hence the α are Hermitian as well. In this formalism for the electron, the wave function has four components, and they all depend on the four x only. Unlike the non-relativistic formalism with spin, one has two extra components which reflect the ability of Eq. (1.1.6) to describe negative-energy states.

By virtue of (1.1.14)–(1.1.17), Eq. (1.1.6) may be re-expressed as

$$\Big[p_0 - \rho_1 \vec{\sigma} \cdot \vec{p} - \rho_3 mc\Big] \psi = 0. \tag{1.1.24}$$

The generalization to the case when an external electromagnetic field is present is not difficult. For this purpose, it is enough to bear in mind that

$$p_b = mv_b + \frac{e}{c} A_b \quad \forall b = 0, 1, 2, 3, \tag{1.1.25}$$

where A_b are such that $A_b dx^b$ is the connection one-form (or 'potential') of the theory. By raising indices with the metric of the background space-time, one obtains the corresponding four-vector

$$A^b = g^{bc}\, A_c. \tag{1.1.26}$$

The desired wave equation of the relativistic theory of the electron in an external electromagnetic field turns out to be

$$\left[p_0 + \frac{e}{c}A_0 - \rho_1\left(\vec{\sigma}, \vec{p} + \frac{e}{c}\vec{A}\right) - \rho_3 mc\right]\psi = 0. \tag{1.1.27}$$

With this notation, the wave function should be regarded as a 'column vector' with four rows, while its 'conjugate imaginary', say $\overline{\psi}^\dagger$, is a row vector, i.e. a 1×4 matrix, and obeys the equation

$$\overline{\psi}^\dagger\left[p_0 + \frac{e}{c}A_0 - \rho_1\left(\vec{\sigma}, \vec{p} + \frac{e}{c}\vec{A}\right) - \rho_3 mc\right] = 0, \tag{1.1.28}$$

where the momentum operators operate to the right.

In the literature, bearing in mind (1.1.13), it is standard practice to define the γ-matrices by means of the anti-commutation property

$$\gamma^a\gamma^b + \gamma^b\gamma^a = 2g^{ab}. \tag{1.1.29}$$

This holds *in any* space-time dimension. In the *Dirac representation*, in four dimensions, one has

$$\gamma^0 = \rho_3, \tag{1.1.30}$$

$$\gamma^i = \gamma^0\, \alpha^i \;\; \forall i = 1, 2, 3, \tag{1.1.31}$$

where the matrices α^i coincide with the ones defined in (1.1.14)–(1.1.16) (the position of indices should not confuse the reader, at this stage).

Denoting by I_2 the 2×2 matrix

$$I_2 \equiv \begin{pmatrix} 1 & 0 \\ 0 & 1 \end{pmatrix},$$

and by τ_2 the second Pauli matrix

$$\tau_2 \equiv \begin{pmatrix} 0 & i \\ -i & 0 \end{pmatrix},$$

one can define two additional representations for the γ-matrices which are quite useful. In the *Majorana representation*, one has (e.g. Itzykson and Zuber 1985)

$$\gamma^a_{\text{Majorana}} = U\, \gamma^a_{\text{Dirac}}\, U^\dagger, \tag{1.1.32}$$

where

$$U \equiv \frac{1}{\sqrt{2}}\begin{pmatrix} I_2 & -\tau_2 \\ -\tau_2 & -I_2 \end{pmatrix} = U^\dagger. \tag{1.1.33}$$

In the *chiral representation*, one has instead

$$\gamma^a_{\text{chiral}} = V \, \gamma^a_{\text{Dirac}} \, V^\dagger, \tag{1.1.34}$$

with

$$V \equiv \frac{1}{\sqrt{2}} \begin{pmatrix} I_2 & -I_2 \\ I_2 & I_2 \end{pmatrix}. \tag{1.1.35}$$

Note that, strictly, we have studied so far the γ-matrices in four-dimensional Minkowski space-time, which is a flat Lorentzian four-manifold. When the Riemann curvature does not vanish, the property (1.1.29) remains a good starting point in the formalism of γ-matrices, but the details are more involved. Some foundational problems in the theory of spinor fields in Lorentzian manifolds are hence described in the following section.

1.2 Spinor fields in Lorentzian manifolds

A four-dimensional space-time (M, g) consists, by definition, of a connected, four-dimensional, Hausdorff C^∞ manifold M, jointly with a Lorentz metric g on M, and a time orientation given by a globally defined timelike vector field $X : M \rightarrow TM$. Indeed, two Lorentzian metrics on M are identified if they are related by the action of the diffeomorphism group of M. One is thus dealing with equivalence classes of pairs (M, g) as a mathematical model of the space-time manifold (Hawking and Ellis 1973). If a Levi–Civita connection, say ∇, is given on M, the corresponding Riemann curvature tensor is defined by

$$R(X, Y)Z \equiv \nabla_X \nabla_Y Z - \nabla_Y \nabla_X Z - \nabla_{[X,Y]} Z. \tag{1.2.1}$$

A curved space-time is a space-time manifold whose curvature tensor does not vanish, and represents, according to Einstein's general relativity, a good mathematical model of the world we live in (Hawking and Ellis 1973). We are now aiming to give an elementary introduction to the theory of spinor fields in a curved space-time, as a first step towards the theory of spinor fields in Riemannian manifolds (whose metric is positive-definite, instead of being Lorentzian).

For this purpose, recall that a *vierbein* is a collection of four oriented mutually orthogonal vectors $e^{\hat{a}}$, $\hat{a} = 0, 1, 2, 3$, whose components $e_a^{\ \hat{a}}$ are related to the metric tensor g_{ab} by

$$g_{ab}(x) = e_a^{\ \hat{c}}(x)\, e_b^{\ \hat{d}}(x)\, \eta_{\hat{c}\hat{d}}, \tag{1.2.2}$$

where $\eta_{\hat{c}\hat{d}} = \mathrm{diag}(1, -1, -1, -1)$ is the Minkowski metric. The relation (1.2.2) is clearly preserved if the vierbein is replaced by its rotated form (Isham 1978):

$$e_a^{\ \hat{c}}(x) \rightarrow e_a^{\ \hat{b}}(x)\, \Omega_{\hat{b}}^{\ \hat{c}}(x), \tag{1.2.3}$$

where $\Omega_{\hat{b}}^{\ \hat{c}}$ is an *arbitrary* differentiable function on space-time, taking values in the group $SO(3,1)$. By definition, spinor fields in curved space-time are required to transform as Dirac spinors under this local gauge group. Thus, one *assumes* that $SL(2,C)$-valued functions exist, say S, such that

$$\Lambda(S(x)) = \Omega(x), \tag{1.2.4}$$

where Λ is the two-to-one covering map of $SL(2,C)$ onto $SO(3,1)$. In the Dirac representation of $SL(2,C)$, one has the matrix relation

$$S^{-1}\, \gamma_{\hat{a}}\, S = \gamma_{\hat{b}}\, \Omega_{\hat{a}}^{\ \hat{b}}, \tag{1.2.5}$$

and the spinor fields transform as

$$\psi(x) \rightarrow S(x)\, \psi(x). \tag{1.2.6}$$

Covariant derivatives of spinor fields are defined according to the rules

$$\nabla_a \psi \equiv \Big(\partial_a + iB_a\Big)\psi, \tag{1.2.7}$$

$$\nabla_a \overline{\psi} \equiv \partial_a \overline{\psi} - i\overline{\psi} B_a, \tag{1.2.8}$$

where B_a, the *spin-connection*, has the transformation property

$$B_a \rightarrow S B_a S^{-1} + i\Big(\partial_a S\Big) S^{-1}, \tag{1.2.9}$$

and is defined by

$$B_a \equiv \frac{1}{4} B_{a\hat{c}\hat{d}} \left[\gamma^{\hat{c}}, \gamma^{\hat{d}}\right]. \tag{1.2.10}$$

A suitable form of the components $B_{a\hat{c}\hat{d}}$ is obtained from Christoffel symbols according to the rule

$$B_{a\hat{c}\hat{d}} = \Gamma^{c}_{\ ab}\, e_{\hat{c}}^{\ b}\, e_{\hat{d}c} + e_{\hat{d}b}\, e^{b}_{\ \hat{c},a}. \tag{1.2.11}$$

By virtue of (1.2.7)–(1.2.11), the gauge-invariant Lagrangian for a (massive) Dirac field is

$$\mathcal{L} = (\det e) \left[\frac{1}{2} \left(\overline{\psi} \, \gamma_{\hat{c}} \, \nabla_a \psi - \left(\nabla_a \overline{\psi} \right) \gamma_{\hat{c}} \, \psi \right) e^{a\hat{c}} - m \overline{\psi} \psi \right]. \quad (1.2.12)$$

Some remarks are now in order (Isham 1978).

(i) Equation (1.2.4) does not hold, in general. The possibility to lift the arbitrary function Ω from $SO(3,1)$, which is not simply connected, to its simply connected covering group $SL(2,C)$, depends on the global topological structure of M. More precisely, a necessary and sufficient condition for the existence of the $SL(2,C)$-valued function S occurring in Eq. (1.2.4) is that (Spanier 1966)

$$\Omega_* \, \pi_1(M) = 0, \quad (1.2.13)$$

where $\pi_1(M)$ is the first homotopy group of space-time. The meaning of Eq. (1.2.13) is that the $SO(3,1)$ vierbein rotation Ω can be covered by a spinor gauge transformation if and only if *any* circle in M has an image in $SO(3,1)$ that can be contracted to a point. Indeed, if

$$\rho : \theta \in [0, 2\pi] \rightarrow \rho(\theta) \quad (1.2.14)$$

is a circle map in M (i.e. $\rho(0) = \rho(2\pi)$), then

$$\Omega : \theta \in [0, 2\pi] \rightarrow \Omega(\rho(\theta)) \quad (1.2.15)$$

is a circle map in $SO(3,1)$, and if ρ belongs to the equivalence class $[\rho] \in \pi_1(M)$ of homotopically related maps, then

$$\Omega_*[\rho]$$

denotes the homotopy class, in $\pi_1(SO(3,1))$, of the map (1.2.15). Thus, if M is not simply connected, there may be functions Ω for which the condition (1.2.13) is not respected.

(ii) The spin-connections corresponding to non-liftable vierbein rotations Ω are obtained by gauge rotations similar, in form, to (1.2.9), but with Ω replacing S. Thus, the gauge rotation can no longer be cancelled by a compensating transformation of spinor fields. This leads in turn to a number of possible Lagrangians, each of which is $SL(2,C)$ gauge-invariant, where the derivative terms in (1.2.12) are replaced by

$$\nabla_a \psi \rightarrow \left(\partial_a + i \Omega B_a \Omega^{-1} - \Omega \partial_a \Omega^{-1} \right) \psi, \quad (1.2.16)$$

$$\nabla_a \overline{\psi} \rightarrow \left(\partial_a - i\Omega B_a \Omega^{-1} + \Omega \partial_a \Omega^{-1}\right) \overline{\psi}, \tag{1.2.17}$$

and $e^{a\hat{c}}$ is replaced by $i\Omega^{\hat{c}}_{\ \hat{d}}\ e^{a\hat{d}}$, with Ω ranging over the equivalence classes of gauge functions. The number of different Lagrangians turns out to coincide with the number of elements in the cohomology group $H^1(M; Z_2)$. This is quite relevant for the path-integral approach to quantum gravity (cf. section 4.4), where one might have to 'sum' over inequivalent vierbein gauge classes, as well as over metric tensors.

A spinor structure in a Lorentzian four-manifold can be defined as a principal $SL(2,C)$ bundle, say E, jointly with a bundle map f from E onto the principal $SO(3,1)$ bundle ξ of oriented orthonormal frames (the vierbeins of the literature on general relativity). The map f has to be compatible with both the $SL(2,C)$ and $SO(3,1)$ actions, in that

$$f(pA) = f(p)\Lambda(A), \tag{1.2.18}$$

where p is *any* point in the bundle space of E, A is any $SL(2,C)$ group element, and pA is the spin-frame into which p is taken under the action of A. Moreover, $f(p)\Lambda(A)$ is the orthonormal frame into which $f(p)$ is taken by the $SO(3,1)$ element $\Lambda(A)$. *Spinor fields* are then the cross-sections of the vector bundle associated with the Dirac representation of $SL(2,C)$ on C^4. If spin-structures do not exist, one can, however, resort to the formalism of $Spin^c$ structures (see Avis and Isham 1980, and our section 6.7).

1.3 Spin-structures: Riemannian case

In the course of defining spin-structures on Riemannian manifolds, we want to give a precise formulation of the following basic ideas. One is given an oriented Riemannian manifold, say M. The tangent bundle of M, $T(M)$, has the rotation group $SO(n)$ as structural group. If it is possible to replace $SO(n)$ by its two-fold covering group $Spin(n)$ (cf. appendix 1.A for the case $n = 3$), one then says that M can be given a spin-structure.

Following Milnor (1963b) we start from a principal bundle, say ξ, over a Riemannian manifold B, with total space denoted by $E(\xi)$. The structural group of ξ is assumed to be $SO(n)$, where n can be *any* positive integer, including $n = \infty$. By definition, a *spin-structure* on ξ is a pair (η, f) consisting of

(i) a principal bundle η over B, with total space $E(\eta)$ and structural group coinciding with $Spin(n)$, i.e. the double covering of $SO(n)$;

(ii) a map $f : E(\eta) \rightarrow E(\xi)$ such that

$$f\,\rho_1 = \rho_2(f \times \lambda), \tag{1.3.1}$$

where λ is the homomorphism from $Spin(n)$ to $SO(n)$, ρ_1 is the right action of $Spin(n)$ and ρ_2 is the right action of $SO(n)$. In other words, there exist two equivalent ways to reach the total space of the given principal bundle ξ over B. First, one goes from $E(\eta) \times Spin(n)$ to $E(\eta)$ via right translation, then using the projection map from $E(\eta)$ to $E(\xi)$. However, one can also use the projection map $f \times \lambda$ from $E(\eta) \times Spin(n)$ to $E(\xi) \times SO(n)$. At this stage, one takes advantage of right translation to go from $E(\xi) \times SO(n)$ to the total space of ξ. A naturally occurring question is how to identify (or distinguish among) spin-structures. The answer is contained in the definition according to which a second spin-structure on ξ, say (η', f') should be identified with (η, f) if there exists an isomorphism h from η' to η such that $f \cdot h = f'$.

In the particular case when $n = 2$, one deals with $Spin(2)$, the two-fold covering group of the circle $SO(2)$; when $n = 1$, one deals with $Spin(1)$, a cyclic group of order 2. Useful examples are given by the tangent bundle of the two-sphere S^2. This has a unique spin-structure (η, f), where $E(\eta)$ is a three-sphere. The tangent bundle of the circle S^1 has instead two distinct spin-structures.

If the reader is more familiar with (or prefers) cohomology classes, he may follow Hirsch and Milnor by giving the following alternative definition.

A spin-structure on ξ is a cohomology class $\sigma \in H^1\Big(E(\xi); Z_2\Big)$ (here, of course, Z_2 denotes the integers modulo 2), whose restriction to each fibre is a generator of the cyclic group $H^1(\text{Fibre}; Z_2)$.

In other words, the underlying idea is that each cohomology class belonging to the group $H^1\Big(E(\xi); Z_2\Big)$ determines a two-fold covering of $E(\xi)$. It is precisely this two-fold covering space that should be taken as the total space $E(\eta)$. The condition involving the restriction to each fibre ensures that each fibre is covered by a copy of $Spin(n)$, i.e. the unique two-fold covering of $SO(n)$. It can be shown that the two definitions are completely equivalent, and hence one is free to choose between the two.

Remarkably, a necessary and sufficient condition for an $SO(n)$ bundle to be endowed with a spin-structure is that its second Stiefel–Whitney class (Milnor and Stasheff 1974), $w_2(\xi)$, should vanish. Further to this, if $w_2(\xi) = 0$, the number of distinct spin-structures on ξ is equal to the number of elements in $H^1(B; Z_2)$. Indeed, if B is connected, this property follows from the exact sequence

$$0 \rightarrow H^1(B; Z_2) \rightarrow H^1(E(\xi); Z_2) \rightarrow H^1(SO(n); Z_2) \rightarrow H^2(B; Z_2),$$

which can be extracted, in turn, from the spectral sequence of the fibration ξ (Milnor 1963b).

By definition, a *spin-manifold* consists of an oriented Riemannian manifold M, jointly with a spin-structure on $T(M)$, the tangent bundle of M. More explicitly, if $F(M)$ denotes the space of oriented orthonormal n-frames on M, one can think of the spin-structure as being a cohomology class $\sigma \in H^1(F(M); Z_2)$, whose restriction to each fibre is non-trivial (for all integer values of n greater than 1). The resulting spin-manifold is then denoted by (M, σ). In particular, if M is simply connected, so that σ is uniquely determined, one simply says that M itself is a spin-manifold.

Dirac spinor fields on the n-dimensional manifold M^n are defined to be *cross-sections* of the vector bundle E^τ over M^n with typical fibre $C^{2[n/2]}$ and structure group $Spin(n)$, where τ is the fundamental spinor representation of $Spin(n)$ (see section 1.4).

When one makes explicit calculations, it is necessary to express a spinor field and its covariant derivative with respect to a (local) orthonormal frame $\{e_a\}_{a=1,\ldots,n}$ on M^n in components. Indeed, a metric connection ω induces a spin-connection on M^n, i.e. a connection on the spin-structure $Spin(M^n)$, and a covariant derivative for the cross-sections. Given a spinor, say ψ, the components of the spinor $\nabla_{e_a}\psi$ in a set of linearly independent local cross-sections of E^τ (such a set is called a local spinor frame, and is denoted by $\{\theta_A\}$) are given by

$$\left(\nabla_{e_a}\psi\right)^A = e_a(\psi^A) - \frac{1}{2}\omega_{abc}\left(\Sigma^{bc}\right)^A{}_B \psi^B, \qquad (1.3.2)$$

where ω_{abc} are the connection coefficients in the frame e_a. As is well known, such coefficients are defined by the condition

$$\nabla_{e_a} e_b = \omega_{abc} e_c. \qquad (1.3.3)$$

The matrices Σ^{ab} are the generators of the representation τ of $Spin(n)$. They can be expressed by the equation (cf. (1.2.10))

$$\Sigma^{ab} = \frac{1}{4}\Big[\Gamma^a, \Gamma^b\Big], \qquad (1.3.4)$$

where Γ^a are Dirac matrices satisfying (cf. (1.1.29))

$$\Gamma^a \, \Gamma^b + \Gamma^b \, \Gamma^a = 2\delta^{ab} \mathbb{1}. \qquad (1.3.5)$$

With this notation, $\mathbb{1}$ is the unit matrix, the indices a, b range from 1 through n, and the Dirac matrices should be thought of as carrying yet further indices A, B:

$$\Gamma^a = \Big(\Gamma^a\Big)^A_{\ B},$$

where A and B range from 1 through $2^{[n/2]}$. The action of the Dirac operator on a spinor is then defined by the equation

$$\nabla \psi = \Gamma^a \nabla_{e_a} \psi. \qquad (1.3.6)$$

At this stage, it is appropriate to build a global theory of the Dirac operator in the elliptic case. This is performed in the following section (cf. section 2.4).

1.4 Global theory of the Dirac operator

In Riemannian four-geometries, the *total* Dirac operator may be defined as a first-order elliptic operator mapping smooth sections of a complex vector bundle to smooth sections of the same bundle. Its action on the sections (i.e. the spinor fields) is given by composition of Clifford multiplication with covariant differentiation. To understand these statements, we first summarize the properties of connections on complex vector bundles, and we then use the basic properties of spin-structures which enable one to understand how to construct the vector bundle relevant for the theory of the Dirac operator.

A complex vector bundle (e.g. Chern 1979) is a bundle whose fibres are isomorphic to complex vector spaces. Denoting by E the total space, by M the base space, one has the projection map $\pi : E \rightarrow M$ and the sections $s : M \rightarrow E$ such that the composition of π with s yields the identity on the base space:

$\pi \cdot s = \mathrm{id}_M$. The sections s represent the physical fields in our applications. Moreover, denoting by T and T^* the tangent and cotangent bundles of M respectively, a connection ∇ is a map from the space of smooth sections of E, say $\Gamma(E)$ (also denoted by $C^\infty(E, M)$), to the space of smooth sections of the tensor-product bundle $\Gamma(T^* \otimes E)$:

$$\nabla : \Gamma(E) \rightarrow \Gamma(T^* \otimes E),$$

such that the following properties hold:

$$\nabla(s_1 + s_2) = \nabla s_1 + \nabla s_2, \tag{1.4.1}$$

$$\nabla(fs) = df \otimes s + f \nabla s, \tag{1.4.2}$$

where $s_1, s_2, s \in \Gamma(E)$ and f is any C^∞ function. The action of the connection ∇ is expressed in terms of the connection matrix θ as

$$\nabla s = \theta \otimes s. \tag{1.4.3}$$

If one takes a section s' related to s by

$$s' = h \, s, \tag{1.4.4}$$

in the light of (1.4.2)–(1.4.4) one finds by comparison that

$$\theta' h = d \, h + h \, \theta. \tag{1.4.5}$$

Moreover, the transformation law of the curvature matrix

$$\Omega \equiv d\theta - \theta \wedge \theta, \tag{1.4.6}$$

is found to be

$$\Omega' = h \, \Omega \, h^{-1}. \tag{1.4.7}$$

We can now describe in more detail the spin-structures and the corresponding complex vector bundle acted upon by the total Dirac operator. Let M be a compact oriented differentiable n-dimensional manifold (without boundary) on which a Riemannian metric is introduced. Let Q be the principal tangential $SO(n)$-bundle of M. As we know from section 1.3, a spin-structure of M is a principal $Spin(n)$-bundle P over M together with a covering map $\tilde{\pi} : P \rightarrow Q$ such that the following commutative structure exists (see figure 1.1). Given the Cartesian product $P \times Spin(n)$, one first reaches P by the right action of $Spin(n)$ on P, and one eventually arrives at Q by the projection map $\tilde{\pi}$. This is equivalent to first reaching the Cartesian product $Q \times SO(n)$ by the map $\tilde{\pi} \times \rho$, and eventually arriving at Q by the right action

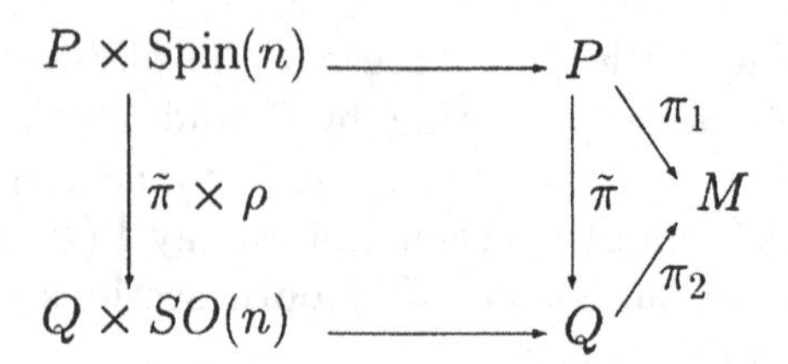

Fig. 1.1. Spin-structure in the Riemannian case.

of $SO(n)$ on Q. Of course, by ρ we denote the double covering $Spin(n) \rightarrow SO(n)$. In other words, P and Q as above are principal fibre bundles over M, and one has a commutative diagram with $P \times Spin(n)$ and P on the top, $Q \times SO(n)$ and Q on the bottom. The projection map from $P \times Spin(n)$ to $Q \times SO(n)$ is $\tilde{\pi} \times \rho$, and the projection map from P to Q is $\tilde{\pi}$. Horizontal arrows should be drawn to denote the right action of $Spin(n)$ on P on the top, and of $SO(n)$ on Q on the bottom. This coincides with the picture resulting from Eq. (1.3.1), apart from the different notation used therein.

The group $Spin(n)$ has a complex representation space Σ of dimension 2^n called the spin-representation. If $G \in Spin(n), x \in \Re^n, u \in \Sigma$, one has therefore

$$G(xu) = GxG^{-1} \cdot G(u) = \rho(G)x \cdot G(u), \tag{1.4.8}$$

where $\rho : Spin(n) \rightarrow SO(n)$ is the covering map as we said before. If M is even-dimensional, i.e. $n = 2l$, the representation is the direct sum of two irreducible representations $\Sigma^{\pm}$ of dimension 2^{n-1}. If M is a $Spin(2l)$ manifold with principal bundle P, one can form the associated complex vector bundles

$$E^+ \equiv P \times \Sigma^+, \tag{1.4.9}$$

$$E^- \equiv P \times \Sigma^-, \tag{1.4.10}$$

$$E \equiv E^+ \oplus E^-. \tag{1.4.11}$$

Sections of these vector bundles are spinor fields on M.

The *total* Dirac operator is a first-order elliptic differential operator $D : \Gamma(E) \rightarrow \Gamma(E)$ defined as follows. Recall first that the Riemannian metric defines a natural $SO(2l)$ connection, and this may be used to give a connection for P. One may therefore consider the connection ∇ at the beginning of this section, i.e. a linear map from $\Gamma(E)$ to $\Gamma(T^* \otimes E)$. On the other hand, the tangent and cotangent bundles of M are isomorphic, and one has the map from $\Gamma(T \otimes E) \rightarrow \Gamma(E)$ induced by *Clifford multiplication* (see section 2.2). The total Dirac operator D is defined to be the *composition* of these two maps. Thus, in terms of an orthonormal base e_i of T, one has *locally*

$$Ds = \sum_i e_i(\nabla_i s), \tag{1.4.12}$$

where $\nabla_i s$ is the covariant derivative of $s \in \Gamma(E)$ in the direction e_i, and $e_i(\ )$ denotes Clifford multiplication. Moreover, the total Dirac operator D induces two operators

$$D^+ : \Gamma(E^+) \rightarrow \Gamma(E^-), \tag{1.4.13}$$

$$D^- : \Gamma(E^-) \rightarrow \Gamma(E^+), \tag{1.4.14}$$

each of which is elliptic. It should be emphasized that ellipticity of the total and partial Dirac operators only holds in Riemannian manifolds, whereas it does not apply to the Lorentzian manifolds of general relativity and of the original Dirac's theory of spin-$\frac{1}{2}$ particles (cf. sections 1.1 and 1.2).

Appendix 1.A

This appendix describes in detail the simplest example of two-to-one homomorphism which is relevant for the theory of spin and spin-structures. Our presentation follows closely the one in Wigner (1959), which relies in turn on a method suggested by H. Weyl. We begin with some elementary results in the theory of matrices, which turn out to be very useful for our purposes.

(i) A matrix which transforms every real vector into a real vector is itself real, i.e. all its elements are real. If this matrix is applied to the kth unit vector (which has kth component $= 1$, all others vanishing), the result is the vector which forms the kth row of the matrix. Thus, this row must be real. But this argument can be applied to all k, and hence all the rows of the matrix must be real.

(ii) It is also well known that a matrix $\mathcal{O}$ is complex orthogonal if it preserves the scalar product of two arbitrary vectors, i.e. if

$$(\vec{a}, \vec{b}) = (\mathcal{O}\vec{a}, \mathcal{O}\vec{b}). \tag{1.A.1}$$

An equivalent condition can be stated in terms of one arbitrary vector: a matrix $\mathcal{O}$ is complex orthogonal if the length of every single arbitrary vector, say $\vec{v}$, is left unchanged under transformation by $\mathcal{O}$. Consider now two arbitrary vectors $\vec{a}$ and $\vec{b}$, and write $\vec{v} = \vec{a} + \vec{b}$. Then our condition for the complex orthogonality of $\mathcal{O}$ is

$$(\vec{v}, \vec{v}) = (\mathcal{O}\vec{v}, \mathcal{O}\vec{v}). \tag{1.A.2}$$

By virtue of the symmetry of the scalar product: $(\vec{a}, \vec{b}) = (\vec{b}, \vec{a})$, this yields

$$\begin{aligned}(\vec{a} + \vec{b}, \vec{a} + \vec{b}) &= (\vec{a}, \vec{a}) + (\vec{b}, \vec{b}) + 2(\vec{a}, \vec{b}) \\ &= (\mathcal{O}\vec{a}, \mathcal{O}\vec{a}) + (\mathcal{O}\vec{b}, \mathcal{O}\vec{b}) + 2(\mathcal{O}\vec{a}, \mathcal{O}\vec{b}).\end{aligned} \tag{1.A.3}$$

However, complex orthogonality also implies that

$$(\vec{a}, \vec{a}) = (\mathcal{O}\vec{a}, \mathcal{O}\vec{a}), (\vec{b}, \vec{b}) = (\mathcal{O}\vec{b}, \mathcal{O}\vec{b}).$$

It then follows from (1.A.3) that

$$(\vec{a}, \vec{b}) = (\mathcal{O}\vec{a}, \mathcal{O}\vec{b}), \tag{1.A.4}$$

which implies that $\mathcal{O}$ is complex orthogonal. It can be shown, in a similar way, that $\mathcal{U}$ is unitary if only $(\vec{v}, \vec{v}) = (\mathcal{U}\vec{v}, \mathcal{U}\vec{v})$ holds for every vector.

By definition, a matrix which leaves each real vector real, and leaves the length of every vector unchanged, is a *rotation.* Indeed, when all lengths are equal in the original and transformed figures, the angles also must be equal; hence the transformation is merely a rotation.

(iii) We now want to determine the general form of a two-dimensional unitary matrix

$$\mathbf{u} = \begin{pmatrix} a & b \\ c & d \end{pmatrix} \tag{1.A.5}$$

of determinant $+1$ by considering the elements of the product

$$\mathbf{u}\mathbf{u}^\dagger = \mathbb{1}. \tag{1.A.6}$$

Recall that the $\dagger$ operation means taking the complex conjugate and then the transposed of the original matrix. Thus, the condition (1.A.6) implies that

$$a^* c + b^* d = 0, \tag{1.A.7}$$

which leads to $c = -b^* d/a^*$. The insertion of this result into the condition of unit determinant:

$$ad - bc = 1, \tag{1.A.8}$$

yields $\left(aa^* + bb^*\right)d/a^* = 1$. Moreover, since $aa^* + bb^* = 1$ from (1.A.6), it follows that $d = a^*$ and $c = -b^*$. The general two-dimensional unitary matrix with unit determinant is hence

$$\mathbf{u} = \begin{pmatrix} a & b \\ -b^* & a^* \end{pmatrix}, \tag{1.A.9}$$

where, of course, we still have to require that $aa^* + bb^* = 1$. Note that, if one writes $a = y_0 + iy_3$ and $b = y_1 + iy_2$, one finds

$$\det \mathbf{u} = y_0^2 + y_1^2 + y_2^2 + y_3^2 = 1.$$

This is the equation of a unit three-sphere centred at the origin, which means that $SU(2)$ has three-sphere topology and is hence simply connected (the n-sphere is simply connected for all $n > 1$). More precisely, $SU(2)$ is homeomorphic to $S^3 \subset \Re^4$.

Consider now the so-called *Pauli matrices*:

$$\tau_x = \begin{pmatrix} 0 & 1 \\ 1 & 0 \end{pmatrix}, \tag{1.A.10}$$

$$\tau_y = \begin{pmatrix} 0 & i \\ -i & 0 \end{pmatrix}, \tag{1.A.11}$$

$$\tau_z = \begin{pmatrix} -1 & 0 \\ 0 & 1 \end{pmatrix}. \tag{1.A.12}$$

Every two-dimensional matrix with zero trace, say $\mathbf{h}$, can be expressed as a linear combination of these matrices:

$$\mathbf{h} = x\tau_x + y\tau_y + z\tau_z = (r, \tau). \tag{1.A.13}$$

Explicitly, one has

$$\mathbf{h} = \begin{pmatrix} -z & x + iy \\ x - iy & z \end{pmatrix}. \tag{1.A.14}$$

In particular, if x, y, and z are real, then $\mathbf{h}$ is Hermitian.

If one transforms $\mathbf{h}$ by an arbitrary unitary matrix $\mathbf{u}$ with unit determinant, one again obtains a matrix with zero trace, $\overline{\mathbf{h}} = \mathbf{u}\mathbf{h}\mathbf{u}^\dagger$. Thus, $\overline{\mathbf{h}}$ can also be written as a linear combination of τ_x, τ_y, τ_z:

$$\overline{\mathbf{h}} = \mathbf{u}\mathbf{h}\mathbf{u}^\dagger = \mathbf{u}(r, \tau)\mathbf{u}^\dagger = x'\tau_x + y'\tau_y + z'\tau_z = (r', \tau), \tag{1.A.15}$$

$$\begin{pmatrix} a & b \\ -b^* & a^* \end{pmatrix} \begin{pmatrix} -z & x+iy \\ x-iy & z \end{pmatrix} \begin{pmatrix} a^* & -b \\ b^* & a \end{pmatrix}$$
$$= \begin{pmatrix} -z' & x'+iy' \\ x'-iy' & z' \end{pmatrix}. \tag{1.A.16}$$

Equation (1.A.16) determines x', y', z' as linear functions of x, y, z. The transformation R_u which carries $r = (x, y, z)$ into $R_u r = r' = (x', y', z')$ can be found from Eq. (1.A.16). It is

$$x' = \frac{1}{2}\Big(a^2 + a^{*2} - b^2 - b^{*2}\Big)x + \frac{i}{2}\Big(a^2 - a^{*2} + b^2 - b^{*2}\Big)y$$
$$+\Big(a^*b^* + ab\Big)z, \tag{1.A.17}$$

$$y' = \frac{i}{2}\Big(a^{*2} - a^2 + b^2 - b^{*2}\Big)x + \frac{1}{2}\Big(a^2 + a^{*2} + b^2 + b^{*2}\Big)y$$
$$+\, i\Big(a^*b^* - ab\Big)z, \tag{1.A.18}$$

$$z' = -(a^*b + ab^*)x + i(a^*b - ab^*)y + (aa^* - bb^*)z. \tag{1.A.19}$$

The particular form of the matrix R_u does not matter; it is important only that

$$x'^2 + y'^2 + z'^2 = x^2 + y^2 + z^2, \tag{1.A.20}$$

since the determinants of $\overline{\mathbf{h}}$ and $\mathbf{h}$ are equal ($\mathbf{u}$ being an element of $SU(2)$). According to the analysis in (ii), this implies that the transformation R_u must be complex orthogonal. Such a property can also be seen directly from (1.A.17)–(1.A.19).

Moreover, $\overline{\mathbf{h}}$ is Hermitian if $\mathbf{h}$ is. In other words, $r' = (x', y', z')$ is real if $r = (x, y, z)$ is real. This implies, by virtue of (i), that R_u is pure real, as can also be seen directly from (1.A.17)–(1.A.19). Thus, R_u is a rotation: every two-dimensional unitary matrix $\mathbf{u}$ of unit determinant corresponds to a three-dimensional rotation R_u; the correspondence is given by (1.A.15) or (1.A.16).

It should be stressed that the determinant of R_u is $+1$, since as $\mathbf{u}$ is changed continuously into a unit matrix, R_u goes continuously into the three-dimensional unit matrix. If its determinant were -1 at the beginning of this process, it would have to make the jump to $+1$. This is impossible, since the function 'det' is continuous, and hence the matrices with negative determinant cannot be connected to the identity of the group. As a corollary of these properties, R_u is a pure rotation for all $\mathbf{u}$.

The above correspondence is such that the product $\mathbf{qu}$ of two unitary matrices $\mathbf{q}$ and $\mathbf{u}$ corresponds to the product $R_{qu} = R_q \cdot R_u$ of the corresponding rotations. According to (1.A.15), applied to $\mathbf{q}$ instead of $\mathbf{u}$,

$$\mathbf{q}(r, \tau)\mathbf{q}^{\dagger} = \left(R_q r, \tau\right), \tag{1.A.21}$$

and upon transformation with $\mathbf{u}$ this yields

$$\mathbf{uq}(r, \tau)\mathbf{q}^{\dagger}\mathbf{u}^{\dagger} = \mathbf{u}(R_q r, \tau)\mathbf{u}^{\dagger} = (R_u R_q r, \tau) = (R_{uq} r, \tau), \tag{1.A.22}$$

using (1.A.15) again, with $R_q r$ replacing r and $\mathbf{uq}$ replacing $\mathbf{u}$. Thus, a homomorphism exists between the group $SU(2)$ of two-dimensional unitary matrices of determinant +1 and rotations in three dimensions; the correspondence is given by (1.A.15) or (1.A.17)–(1.A.19). Recall, by the way, that a homomorphism of two groups, say G_1 and G_2, is a map $\phi : G_1 \to G_2$ such that

$$\phi(g_1 \cdot g_2) = \phi(g_1)\, \phi(g_2),$$

$$\phi\left(e_{G_1}\right) = e_{G_2},$$

$$\phi(g^{-1}) = (\phi(g))^{-1}.$$

However, we note that so far we have not shown that the homomorphism exists between the two-dimensional unitary group and the *whole* pure rotation group. That would imply that R_u ranges over all rotations as $\mathbf{u}$ covers the entire unitary group. This will be proved shortly. It should also be noticed that the homomorphism *is not an isomorphism*, since more than one unitary matrix corresponds to the same rotation (see below).

We first assume that $\mathbf{u}$ is a diagonal matrix, say $\mathbf{u}_1(\alpha)$ (i.e. we set $b = 0$, and, for convenience, we write $a = e^{-\frac{i}{2}\alpha}$). Then $| a^2 | = 1$ and α is real:

$$\mathbf{u}_1(\alpha) = \begin{pmatrix} e^{-\frac{i}{2}\alpha} & 0 \\ 0 & e^{\frac{i}{2}\alpha} \end{pmatrix}. \tag{1.A.23}$$

From (1.A.17)–(1.A.19) one can see that the corresponding rotation:

$$R_{u_1} = \begin{pmatrix} \cos\alpha & \sin\alpha & 0 \\ -\sin\alpha & \cos\alpha & 0 \\ 0 & 0 & 1 \end{pmatrix} \tag{1.A.24}$$

is a rotation about Z through an angle α. We next assume that

$\mathbf{u}$ is real:

$$\mathbf{u}_2(\beta) = \begin{pmatrix} \cos\frac{\beta}{2} & -\sin\frac{\beta}{2} \\ \sin\frac{\beta}{2} & \cos\frac{\beta}{2} \end{pmatrix}. \tag{1.A.25}$$

From (1.A.17)–(1.A.19) the corresponding rotation is found to be

$$R_{u_2} = \begin{pmatrix} \cos\beta & 0 & -\sin\beta \\ 0 & 1 & 0 \\ \sin\beta & 0 & \cos\beta \end{pmatrix}, \tag{1.A.26}$$

i.e. a rotation about Y through an angle β. The product of the three unitary matrices $\mathbf{u}_1(\alpha)\mathbf{u}_2(\beta)\mathbf{u}_1(\gamma)$ corresponds to the product of a rotation about Z through an angle γ, about Y through β, and about Z through α, in other words, to a rotation with Euler angles α, β, γ. It follows from this that the correspondence defined in (1.A.15) not only specified a three-dimensional rotation for every two-dimensional unitary matrix, but also *at least one unitary matrix* for every pure rotation. Specifically, the matrix

$$\begin{pmatrix} e^{-\frac{i}{2}\alpha} & 0 \\ 0 & e^{\frac{i}{2}\alpha} \end{pmatrix} \begin{pmatrix} \cos\frac{\beta}{2} & -\sin\frac{\beta}{2} \\ \sin\frac{\beta}{2} & \cos\frac{\beta}{2} \end{pmatrix} \begin{pmatrix} e^{-\frac{i}{2}\gamma} & 0 \\ 0 & e^{\frac{i}{2}\gamma} \end{pmatrix}$$

$$= \begin{pmatrix} e^{-\frac{i}{2}\alpha}\cos\frac{\beta}{2}e^{-\frac{i}{2}\gamma} & -e^{-\frac{i}{2}\alpha}\sin\frac{\beta}{2}e^{\frac{i}{2}\gamma} \\ e^{\frac{i}{2}\alpha}\sin\frac{\beta}{2}e^{-\frac{i}{2}\gamma} & e^{\frac{i}{2}\alpha}\cos\frac{\beta}{2}e^{\frac{i}{2}\gamma} \end{pmatrix} \tag{1.A.27}$$

corresponds to the rotation $\{\alpha\beta\gamma\}$. Thus, the homomorphism is in fact a homomorphism of the unitary group onto the *whole* three-dimensional pure rotation group.

The question remains of the *multiplicity* of the homomorphism, i.e. how many unitary matrices $\mathbf{u}$ correspond to the same rotation. For this purpose, it is sufficient to check how many unitary matrices $\mathbf{u}_0$ correspond to the identity of the rotation group, i.e. to the transformation $x' = x, y' = y, z' = z$. For these particular $\mathbf{u}_0$, the identity $\mathbf{u}_0\mathbf{h}\mathbf{u}_0^\dagger = \mathbf{h}$ should hold for all $\mathbf{h}$; this can only be the case when $\mathbf{u}_0$ is a constant matrix: $b = 0$ and $a = a^*$, $\mathbf{u}_0 = \pm\mathbb{I}$ (since $|a|^2 + |b|^2 = 1$). Thus, the two unitary matrices $+\mathbb{I}$ and $-\mathbb{I}$, and *only these*, correspond to the identity of the rotation group. These two elements form an invariant sub-group of the unitary group, and those elements (and only those) which are in the same coset of the invariant sub-group, i.e. $\mathbf{u}$ and $-\mathbf{u}$, correspond to the same rotation. Indeed, that $\mathbf{u}$ and $-\mathbf{u}$ actually do correspond to the same rotation can be seen directly from (1.A.15) or (1.A.17)–(1.A.19).

Alternatively, one can simply note that only the half-Euler angles occur in (1.A.27). The Euler angles are determined by a rotation only up to a multiple of 2π; the half angles, only up to a multiple of π. This implies that the trigonometric functions in (1.A.27) are determined only up to a sign.

A very important result has been thus obtained: there exists a two-to-one homomorphism of the group of two-dimensional unitary matrices with determinant 1 *onto* the three-dimensional pure rotation group: there is a one-to-one correspondence between *pairs* of unitary matrices $\mathbf{u}$ and $-\mathbf{u}$ and rotations R_u in such a way that, from $\mathbf{uq} = \mathbf{t}$ it also follows that $R_u R_q = R_t$; conversely, from $R_u R_q = R_t$, one has that $\mathbf{uq} = \pm\mathbf{t}$. If the unitary matrix $\mathbf{u}$ is known, the corresponding rotation is best obtained from (1.A.17)–(1.A.19). Conversely, the unitary matrix for a rotation $\{\alpha\beta\gamma\}$ is best found from (1.A.27) (Wigner 1959).

2
Differential operators on manifolds

On studying differential operators on manifolds, a key role is played by the operators of Laplace type. These are second-order elliptic operators with leading symbol given by the Riemannian metric of the base space, M. They also include the effect of an endomorphism of the vector bundle over M. As a next step, Clifford algebras are introduced, because they are quite essential to define the Dirac operator. The general case of the Clifford algebra of a vector space equipped with a non-degenerate quadratic form is described. Clifford groups and Pin groups are also briefly considered. The following sections are devoted to the signature operator, an intrinsic definition of ellipticity and index, a brief review of properties of the Dirac operator (first in Euclidean space and then on compact, oriented, even-dimensional Riemannian manifolds), with emphasis on the spectral flow and θ-functions. The chapter ends with an introduction to pseudo-differential operators and non-local boundary-value problems.

2.1 Operators of Laplace type

In the analysis of differential operators on manifolds, the standard framework consists of a vector bundle, say V, over a Riemannian manifold M endowed with a positive-definite metric g. We want to introduce the concept of *operator of Laplace type*. For this purpose, following Branson *et al.* (1997), we use Greek indices to index local *coordinate frames* ∂_ν and dx^ν for the tangent and cotangent bundles $T(M)$ and $T^*(M)$, respectively. They range from 1 through m, where $m = \dim(M)$. By definition, an operator of Laplace type can be decomposed, *locally*, in the form

$$D = -\Big(g^{\mu\nu} I_V \partial_\mu \partial_\nu + a^\rho \partial_\rho + b\Big), \tag{2.1.1}$$

where a is a local section of $TM \otimes End(V)$, and b is a local section of $End(V)$. It is also useful to express D in terms of Christoffel symbols and connection one-forms. For this purpose, let Γ be the Christoffel symbols of the Levi–Civita connection of the metric g on M, let ∇ be an auxiliary connection on V, and let E be an element of $C^\infty(End(V))$. One can now define (Branson *et al.* 1997):

$$P(g, \nabla, E) \equiv -\Big(\mathrm{Tr}_g \nabla^2 + E\Big)$$
$$= -g^{\mu\sigma}\Big[I_V \partial_\mu \partial_\sigma + 2\omega_\mu \partial_\sigma - \Gamma_{\mu\sigma}^{\ \ \nu} I_V \partial_\nu + \partial_\mu \omega_\sigma$$
$$+ \omega_\mu \omega_\sigma - \Gamma_{\mu\sigma}^{\ \ \nu} \omega_\nu\Big] - E. \tag{2.1.2}$$

A very useful lemma states that, if D is an operator of Laplace type, there exists a *unique connection* ∇ on V and a *unique endomorphism* E of V such that (Branson *et al.* 1997)

$$D = P(g, \nabla, E). \tag{2.1.3}$$

If ω is the connection one-form of ∇, one finds

$$\omega_\rho = \frac{1}{2} g_{\nu\rho}\Big(a^\nu + g^{\mu\sigma} \Gamma_{\mu\sigma}^{\ \ \nu} I_V\Big), \tag{2.1.4}$$

and the endomorphism is expressed by

$$E = b - g^{\mu\nu}\Big(\partial_\mu \omega_\nu + \omega_\mu \omega_\nu - \omega_\sigma \Gamma_{\mu\nu}^{\ \ \sigma}\Big). \tag{2.1.5}$$

The Levi–Civita connection of the metric g, jointly with the connection ∇ on V, are used to evaluate covariant derivatives of tensors of all type (see also section 5.2).

This way of expressing operators of Laplace type is the most convenient from the point of view of applications to heat-kernel asymptotics, which is studied in chapters 4, 5 and 6. However, in the purely mathematical literature, one begins by studying the actual *Laplace operator*, which is expressed in terms of exterior differentiation d, and its adjoint d^*. Indeed, if M is an m-dimensional C^∞ manifold, and if $\Omega^q(M)$ denotes the space of C^∞ exterior differential forms of degree q on M, one has a naturally occurring differential operator

$$d : \Omega^q(M) \rightarrow \Omega^{q+1}(M),$$

which extends the differential of a function. The Laplace (or Hodge–Laplace) operator on Ω^q is defined as

$$\Delta \equiv dd^* + d^*d, \tag{2.1.6}$$

and the harmonic forms $\mathcal{H}^q$ are the solutions of $\Delta u = 0$. The main theorem of Hodge theory is that $\mathcal{H}^q$ is isomorphic to the q de Rham group and hence to the qth cohomology group of M.

2.2 Clifford algebras

Let V be a real vector space equipped with an inner product $\langle\ ,\ \rangle$, defined by a non-degenerate quadratic form Q of signature (p, q). Let $T(V)$ be the tensor algebra of V and consider the ideal $\mathcal{I}$ in $T(V)$ generated by $x \otimes x + Q(x)$. By definition, $\mathcal{I}$ consists of sums of terms of the kind $a \otimes \left\{ x \otimes x + Q(x) \right\} \otimes b$, $x \in V, a, b \in T(V)$. The quotient space

$$Cl(V) \equiv Cl(V, Q) \equiv T(V)/\mathcal{I} \tag{2.2.1}$$

is the *Clifford algebra* of the vector space V equipped with the quadratic form Q. The product induced by the tensor product in $T(V)$ is known as Clifford multiplication or the Clifford product and is denoted by $x \cdot y$, for $x, y \in Cl(V)$. The dimension of $Cl(V)$ is 2^n if $\dim(V) = n$. A basis for $Cl(V)$ is given by the scalar 1 and the products

$$e_{i_1} \cdot e_{i_2} \cdot e_{i_n} \quad i_1 < \ldots < i_n,$$

where $\left\{ e_1, \ldots, e_n \right\}$ is an orthonormal basis for V. Moreover, the products satisfy (cf. (1.1.13))

$$e_i \cdot e_j + e_j \cdot e_i = 0 \quad i \neq j, \tag{2.2.2}$$

$$e_i \cdot e_i = -2\langle e_i, e_i \rangle \quad i = 1, \ldots, n. \tag{2.2.3}$$

As a vector space, $Cl(V)$ is isomorphic to $\Lambda^*(V)$, the Grassmann algebra, with

$$e_{i_1} \ldots e_{i_n} \longrightarrow e_{i_1} \wedge \ldots \wedge e_{i_n}.$$

There are two natural *involutions* on $Cl(V)$. The first, denoted by $\alpha : Cl(V) \rightarrow Cl(V)$, is induced by the involution $x \rightarrow -x$ defined on V, which extends to an automorphism of $Cl(V)$. The eigenspace of α with eigenvalue $+1$ consists of the even elements of $Cl(V)$, and the eigenspace of α of eigenvalue -1 consists of the odd elements of $Cl(V)$.

The second involution is a map $x \rightarrow x^t$, induced on generators by

$$\left(e_{i_1}...e_{i_p}\right)^t = e_{i_p}...e_{i_1},$$

where e_i are basis elements of V. Moreover, we define $x \rightarrow \overline{x}$, a third involution of $Cl(V)$, by $\overline{x} \equiv \alpha(x^t)$.

One then defines $Cl^*(V)$ to be the group of invertible elements of $Cl(V)$, and the *Clifford group* $\Gamma(V)$ is the sub-group of $Cl^*(V)$ defined by

$$\Gamma(V) \equiv \left\{x \in Cl^*(V) : y \in V \Rightarrow \alpha(x) y x^{-1} \in V\right\}. \tag{2.2.4}$$

One can show that the map $\rho : V \rightarrow V$ given by $\rho(x)y = \alpha(x)yx^{-1}$ is an isometry of V with respect to the quadratic form Q. The map $x \rightarrow \|x\| \equiv x\overline{x}$ is the square-norm map, and enables one to define a remarkable sub-group of the Clifford group, i.e. (Ward and Wells 1990)

$$\mathrm{Pin}(V) \equiv \left\{x \in \Gamma(V) : \|x\| = 1\right\}. \tag{2.2.5}$$

2.3 Index of elliptic operators and signature operator

It is convenient to consider the operator $d+d^*$ acting on $\Omega^*(M) \equiv \oplus_q \Omega^q(M)$. Since $d^2 = (d^*)^2 = 0$, one has $\triangle = (d+d^*)^2$, and the harmonic forms are also the solutions of $(d+d^*)u = 0$. If we consider $d+d^*$ as an operator $\Omega^{\mathrm{ev}} \rightarrow \Omega^{\mathrm{odd}}$, its null-space is $\oplus \mathcal{H}^{2q}$, while the null-space of its adjoint is $\oplus \mathcal{H}^{2q+1}$. Hence its *index* is the Euler characteristic of M. It is now appropriate to define elliptic differential operators and their index.

Let us denote again by M a compact oriented smooth manifold, and by E, F two smooth complex vector bundles over M. We consider linear differential operators

$$D : \Gamma(E) \rightarrow \Gamma(F), \tag{2.3.1}$$

i.e. linear operators defined on the spaces of smooth sections and expressible locally by a matrix of partial derivatives. Note that the extra generality involved in considering vector bundles presents no serious difficulties and it is quite essential in a systematic treatment on manifolds, since all geometrically interesting operators operate on vector bundles.

Let $T^*(M)$ denote the cotangent vector bundle of M, $S(M)$ the unit sphere-bundle in $T^*(M)$ (relative to some Riemannian metric), $\pi : S(M) \rightarrow M$ the projection. Then, associated with D there is a vector-bundle homomorphism

$$\sigma(D) : \pi^* E \rightarrow \pi^* F, \tag{2.3.2}$$

which is called the *leading symbol* of D. In terms of local coordinates, $\sigma(D)$ is obtained from D replacing $\frac{\partial}{\partial x_j}$ by $i\xi_j$ in the highest-order terms of D (ξ_j is the jth coordinate in the cotangent bundle). By definition, D is *elliptic* if $\sigma(D)$ is an *isomorphism.* Of course, this implies that the complex vector bundles E, F have the same dimension.

One of the basic properties of elliptic operators is that Ker D (i.e. the null-space) and Coker $D \equiv \Gamma(F)/D\Gamma(E)$ are both finite-dimensional. The *index* $\gamma(D)$ is defined by (Atiyah and Singer 1963)

$$\gamma(D) \equiv \text{dim Ker } D - \text{dim Coker } D. \tag{2.3.3}$$

If $D^* : \Gamma(F) \rightarrow \Gamma(E)$ denotes the formal adjoint of D, relative to metrics in E, F, M, then D^* is also elliptic and

$$\text{Coker } D \cong \text{Ker } D^*, \tag{2.3.4}$$

so that

$$\gamma(D) = \text{dim Ker } D - \text{dim Ker } D^*. \tag{2.3.5}$$

Getting back to the definition of the leading symbol, we find it helpful for the reader to say that, for the exterior derivative d, it coincides with exterior multiplication by $i\xi$, for $\triangle$ it is $\|\xi\|^2$, and for $d + d^*$ it is iA_ξ, where A_ξ is Clifford multiplication by ξ.

For a given $\alpha \in \Lambda^p(\Re^{2l}) \otimes_\Re C$ (recall that $\Lambda^q(V)$ denotes the exterior powers of the vector space V), let us denote by τ the involution

$$\tau(\alpha) \equiv i^{p(p-1)+l} \; {}^*\alpha. \tag{2.3.6}$$

If $n = 2l$, τ can be expressed as $i^l \omega$, where ω denotes Clifford multiplication by the volume form ^{*1}. Remarkably, $(d + d^*)$ and τ anti-commute. In the Clifford algebra one has $\xi\omega = -\omega\xi$ for

$\xi \in T^*$. Thus, if $\Omega^{\pm}$ denote the ± 1-eigenspaces of τ, $(d+d^*)$ maps Ω^+ into Ω^-. The restricted operator

$$d + d^* : \Omega_+ \rightarrow \Omega_-, \tag{2.3.7}$$

is called the *signature operator* and denoted by A. If $l = 2k$, the index of this operator is equal to (dim $\mathcal{H}_+$ − dim $\mathcal{H}_-$), where $\mathcal{H}_{\pm}$ are the spaces of harmonic forms in $\Omega_{\pm}$. If $q \neq l$, the space $\mathcal{H}^q \oplus \mathcal{H}^{n-q}$ is stable under τ, which just interchanges the two factors. Hence one gets a vanishing contribution to the index from such a space. For $q = l = 2k$, however, one has $\tau(\alpha) = i^{l(l-1)+l} {*\alpha} = {*\alpha}$ for $\alpha \in \Omega^l$. One thus finds

$$\text{index } A = \dim \mathcal{H}^l_+ - \dim \mathcal{H}^l_-, \tag{2.3.8}$$

where $\mathcal{H}^l_{\pm}$ are the ± 1-eigenspaces of $*$ on $\mathcal{H}^l$. Thus, since the inner product in $\mathcal{H}^l$ is given by $\int u_\wedge {}^*v$, one finds

$$\text{index } A = \text{Sign}(M), \tag{2.3.9}$$

where Sign(M) is the signature of the quadratic form on the cohomology group $H^{2k}(M; \Re)$. Hence the name for the operator A.

It is now worth recalling that the *cup-product* is a very useful algebraic structure occurring in cohomology theory. It works as follows. Given $[\omega] \in H^p(M; \Re)$ and $[\nu] \in H^q(M; \Re)$, one then defines the cup-product of $[\omega]$ and $[\nu]$, written $[\omega] \bigcup [\nu]$, by

$$[\omega] \bigcup [\nu] \equiv [\omega \wedge \nu]. \tag{2.3.10}$$

The right-hand side of (2.3.10) is a $(p+q)$-form so that $[\omega \wedge \nu] \in H^{p+q}(M; \Re)$. One can check, using the properties of closed and exact forms, that this is a well-defined product of cohomology classes. The cup-product $\bigcup$ is therefore a map of the form

$$\bigcup : H^p(M; \Re) \times H^q(M; \Re) \rightarrow H^{p+q}(M; \Re). \tag{2.3.11}$$

If the sum of all cohomology groups, $H^*(M; \Re)$, is defined by

$$H^*(M; \Re) \equiv \oplus_{p>0} H^p(M; \Re), \tag{2.3.12}$$

the cup-product has the neater looking form

$$\bigcup : H^*(M; \Re) \times H^*(M; \Re) \rightarrow H^*(M; \Re). \tag{2.3.13}$$

The product on $H^*(M; \Re)$ defined in (2.3.13) makes $H^*(M; \Re)$ into a ring. It can happen that two spaces M and N have the same cohomology groups and yet are not topologically the same. This can be proved by evaluating the cup-product $\bigcup$ for $H^*(M; \Re)$

and $H^*(N;\Re)$, and showing that the resulting rings are different. An example is provided by choosing $M \equiv S^2 \times S^4$ and $N \equiv CP^3$.

2.4 Dirac operator

We begin our analysis by considering Euclidean space $\Re^{2k}$. The Clifford algebra $\widetilde{C}_{2k}$ of section 2.2 is a full matrix algebra acting on the 2^k-dimensional spin-space S. Moreover, S is given by the direct sum of S^+ and S^-, i.e. the eigenspaces of $\omega = e_1 e_2 ... e_{2k}$, and $\omega = \pm i^k$ on $S^\pm$. Let us now denote by $E_1, ..., E_{2k}$ the linear transformations on S representing $e_1, ..., e_{2k}$. The Dirac operator is then defined as the first-order differential operator

$$D \equiv \sum_{i=1}^{2k} E_i \frac{\partial}{\partial x_i}, \tag{2.4.1}$$

and satisfies $D^2 = \Delta \cdot I$, where I is the identity of spin-space. Note that, since the e_i anti-commute with ω, the Dirac operator interchanges the positive and negative spin-spaces, S^+ and S^- (cf. appendix 4.A). Restriction to S^+ yields the operator

$$B : C^\infty\Big(S^+, \Re^{2k}\Big) \rightarrow C^\infty\Big(S^-, \Re^{2k}\Big). \tag{2.4.2}$$

Since $E_i^2 = -1$, in the standard metric the E_i are unitary and hence skew-adjoint. Moreover, since $\frac{\partial}{\partial x_i}$ is also formally skew-adjoint, it follows that D is formally self-adjoint, i.e. the formal adjoint of B is the restriction of D to S^-.

The next step is the global situation of a compact oriented $2k$-dimensional manifold M. First, we assume that a spin-structure exists on M. From section 1.3, this means that the principal $SO(2k)$ bundle P of M, consisting of oriented orthonormal frames, lifts to a principal $Spin(2k)$ bundle Q, i.e. the map $Q \rightarrow P$ is a double covering inducing the standard covering $Spin(2k) \rightarrow SO(2k)$ on each fibre. If $S^\pm$ are the two half-spin representations of $Spin(2k)$, we consider the associated vector bundles on M: $E^\pm \equiv Q \times_{Spin(2k)} S^\pm$. Sections of these vector bundles are the spinor fields on M. The *total* Dirac operator on M is a differential operator acting on $E = E^+ \oplus E^-$ and switching factors as above. To define D we have to use the Levi–Civita connection on P, which lifts to one on Q. This enables one to define the covariant derivative

$$\nabla : C^\infty\Big(E, M\Big) \to C^\infty\Big(E, M \otimes T^*\Big), \tag{2.4.3}$$

and D is defined as the composition of ∇ with the map $C^\infty\Big(E \otimes T^*, M\Big) \to C^\infty(E, M)$ induced by Clifford multiplication. By using an orthonormal base e_i of T at any point one can write (see (1.4.12))

$$Ds \equiv \sum_{i=1}^{2k} e_i \nabla_i s, \tag{2.4.4}$$

where $\nabla_i s$ is the covariant derivative in the direction e_i, and $e_i(\ )$ denotes Clifford multiplication. By looking at symbols, D is skew-adjoint. Hence $D - D^*$ is an algebraic invariant of the metric and it only involves the first derivatives of g. Using normal coordinates one finds that any such invariant vanishes, which implies $D = D^*$, i.e. D is self-adjoint.

As in Euclidean space, the restriction of D to the half-spinors E^+ is denoted by

$$B : C^\infty\Big(E^+, M\Big) \to C^\infty\Big(E^-, M\Big). \tag{2.4.5}$$

The index of the restricted Dirac operator is given by

$$\text{index } B = \dim \mathcal{H}^+ - \dim \mathcal{H}^-, \tag{2.4.6}$$

where $\mathcal{H}^\pm$ are the spaces of solutions of $Du = 0$, for u a section of $E^\pm$. Since D is elliptic and self-adjoint, these spaces are also the solutions of $D^2 u = 0$, and their elements are called *harmonic spinors* (see Lichnerowicz 1963, Hitchin 1974, Bär and Schmutz 1992, Bär 1995, 1996a, 1997).

The reader should be aware of a fundamental difference between the operators of Laplace type of section 2.1 and the operators of Dirac type (by Dirac type one means a linear combination of the Dirac operator and of a suitable endomorphism). What happens is that the spectrum of the former is contained in a half line because it is bounded from below, whereas the spectrum of the latter lies *on the whole real line.* This property is the underlying reason of the topological effects associated with fermionic fields. What happens is that the half line is topologically uninteresting because it can be pushed to infinity, whereas, in the case of the real line, obstructions result from the points at $+\infty$ and $-\infty$. The image suggested by Atiyah (1984a) to appreciate this difference is the one of a string with two infinite ends, which is topologically equivalent to a circle.

In the investigation of topological effects due to a Dirac operator, it is useful to consider a periodic one-parameter family of Dirac type operators, say D_t (Atiyah 1984a). If the base space M is compact, the energy levels $E_n(t)$ are discrete for each value of the parameter t. As t varies in the closed interval $[0, 1]$, the energy levels change, being functions of t, subject to the condition that, at the end of the period, they should coincide with the original value. In general, one has

$$E_n(1) = E_{n+k}(0), \tag{2.4.7}$$

where the integer k does not necessarily vanish. A possible interpretation of k is that it represents the number of negative energy levels that have become positive. In the mathematical literature, k is called the *spectral flow* of the family of Dirac operators, and its introduction is due to Atiyah and Lusztig. Interestingly, if the family D_t is deformed continuously into another periodic family D'_t of the same type, then D_t and D'_t turn out to have the same spectral flow. Hence one says that the spectral flow is a topological invariant of the family D_t.

As an example, following Atiyah (1984a), we take M to be a one-dimensional manifold, with the family D_t given by (cf. section 4.3)

$$D_t \equiv -i\frac{d}{dx} + t, \quad t \in [0, 1], \tag{2.4.8}$$

with periodic boundary conditions

$$f(x) = f(x + 2\pi). \tag{2.4.9}$$

The solutions of the eigenvalue equation

$$D_t f = \lambda f \tag{2.4.10}$$

are found to be

$$f_t(x) = f_0 e^{i(\lambda - t)x}, \tag{2.4.11}$$

so that the boundary condition (2.4.9) leads to

$$\lambda = E_n(t) = n + t. \tag{2.4.12}$$

By comparison with Eq. (2.4.7) one finds

$$E_n(1) = n + 1 = E_{n+1}(0), \tag{2.4.13}$$

i.e. the spectral flow takes the value 1 in our case. The periodicity of the family D_t means that D_1 and D_0 are gauge equivalent, in that the following operator identity holds:

$$e^{-ix} D_0 e^{ix} = -i\frac{d}{dx} + 1 = D_1. \tag{2.4.14}$$

A generalization of the result (2.4.14) can be obtained by considering the family $D_{kt} \equiv -i\frac{d}{dx} + kt$, for which the Leibniz rule yields

$$e^{-ikt} D_{kt} e^{ikt} = -i\frac{d}{dx} + (k+1)t = D_{(k+1)t}. \tag{2.4.15}$$

The geometric interpretation of these calculations is that the spectral flow is related to the *winding number*, which describes the components of the group of $U(1)$ gauge transformations on the circle.

It is now instructive, motivated by Euclidean field theory, to regard t as the Euclidean-time variable, say τ. Correspondingly, following again Atiyah (1984a), we define the operator $\mathcal{D}$ in the (x,τ) space:

$$\mathcal{D} \equiv \frac{\partial}{\partial \tau} + D_\tau = \frac{\partial}{\partial \tau} + \tau - i\frac{\partial}{\partial x}. \tag{2.4.16}$$

This is viewed as an operator acting on functions of the variables x, τ such that

$$f(x,\tau) = f(x + 2\pi, \tau), \tag{2.4.17}$$

$$f(x, \tau + 1) = e^{-ix} f(x, \tau). \tag{2.4.18}$$

Equation (2.4.17) expresses periodicity (of period 2π) in the x variable in the strict sense, whereas Eq. (2.4.18) is a period-1 condition in the τ variable, up to the appropriate gauge transformation. The equation

$$\mathcal{D} f = 0 \tag{2.4.19}$$

can be solved by separation of variables, i.e. as a sum of elementary solutions of the form $f_1(x) f_2(\tau)$. This leads to the equations

$$\left(\frac{1}{f_2} \frac{df_2}{d\tau} + \tau \right) = \frac{i}{f_1} \frac{df_1}{dx} = \kappa. \tag{2.4.20}$$

At this stage, we remark that the periodicity condition (2.4.17) forces κ to take all integer values from $-\infty$ to $+\infty$. Thus, f_1 is proportional to $\exp(inx)$, and f_2 is proportional to

$$\exp\left(-\frac{1}{2}(\tau+n)^2\right),$$

so that the desired solution of Eq. (2.4.19) reads

$$f(x,\tau)=C\sum_{n=-\infty}^{\infty}\exp\left[-\frac{1}{2}(\tau+n)^2+inx\right], \tag{2.4.21}$$

where C is a constant scalar quantity. Equation (2.4.21) is known to define the θ-function, and is the unique solution of the problem (2.4.17)–(2.4.19).

By contrast, denoting by $\mathcal{D}^*$ the formal adjoint of $\mathcal{D}$, one finds no non-trivial finite solutions of the equation

$$\mathcal{D}^* f = 0, \tag{2.4.22}$$

because in $\mathcal{D}^*$ the sign of $\frac{\partial}{\partial\tau}$ is negative, and hence the part of the solution depending on τ is exponentially increasing, unlike the case of (2.4.21). Thus, by virtue of the formula (2.3.5) for the index, the index of $\mathcal{D}$ turns out to be 1.

Interestingly, the θ-functions can be shown to be sections of a non-trivial complex line-bundle over a two-dimensional torus. The existence of non-trivial zero-modes of the total Dirac operator $\mathcal{D}\oplus\mathcal{D}^*$ points out that one is dealing with a gauge sector having non-trivial instanton number (Atiyah 1984a).

A general framework for the investigation of topological properties of Dirac operators is provided by the mathematical theory of Fredholm operators. Since one wants to make sense of equations of the kind

$$Au = 0, \tag{2.4.23}$$

$$A^* v = 0, \tag{2.4.24}$$

for *bounded operators* in Hilbert space, and having only *finitely many independent solutions*, one can replace the original differential operator, say B, with $B(1+B^*B)^{-1/2}$. Such conditions define indeed the *Fredholm operators*, and the set $\mathcal{F}$ of all Fredholm operators on a Hilbert space has a (natural) topology given by the operator norm. An important subspace of $\mathcal{F}$ consists of the self-adjoint Fredholm operators, and is denoted by $\mathcal{F}^1$. When the base space M is odd-dimensional and compact, the Dirac operator provides, upon rescaling (see above) an operator in $\mathcal{F}^1$. By contrast, when M is even-dimensional, half the Dirac operator may be used to yield an operator in $\mathcal{F}$ (Atiyah 1984a).

2.5 Some outstanding problems

In the mathematical and physical applications of elliptic operators, one studies a mathematical framework given by (complex) vector bundles over a compact Riemannian manifold, say M, where M may or may not have a boundary. The latter case is simpler and came first in the (historical) development of index theory. The investigations in Atiyah and Singer (1968a,b, 1971a,b) and Atiyah and Segal (1968) proved that the index of an elliptic operator (in general, one deals with elliptic systems) is obtained by integrating over the sphere-bundle of M a suitable composition of the Chern character with the Todd class of M (see section 3.1). In other words, a deep relation exists between the dimensions of the null-spaces of an elliptic operator and its adjoint on the one hand, and the topological invariants of the fibre bundles of the problem on the other hand.

The classical Riemann–Roch problem, which is concerned with giving a formula for the dimension of the space of meromorphic functions on a compact Riemann surface, having poles of orders $\leq \nu_i$ at points P_i, may be viewed as an index problem. In other words, the solution of the index problem in general is equivalent to finding an extension of the Riemann–Roch theorem from the domain of holomorphic function theory to that of general elliptic systems (Atiyah 1966). In general, one has to find suitable topological invariants of the pair (M, D), where M is the base manifold and D is the elliptic operator. Moreover, the explicit formula for index(D) has to be expressed in terms of these invariants. In the particular case of the Riemann–Roch theorem, the desired topological invariants turn out to be the genus, and the degree of $\sum_i \nu_i P_i$.

It is now instructive to gain a qualitative understanding of why an elliptic operator, say D, defines invariants of the characteristic class type (Atiyah 1966). For this purpose, let us recall that a homogeneous, constant coefficient, $N \times N$ matrix of differential operators

$$P = \left[P_{ij} \left(\frac{\partial}{\partial x_1}, ..., \frac{\partial}{\partial x_n} \right) \right] \quad , \quad i, j = 1, ..., N, \tag{2.5.1}$$

is elliptic if, for $\xi = (\xi_1, ..., \xi_n) \neq 0$ (and real) one has

$$\det P(\xi_1, ..., \xi_n) \neq 0. \tag{2.5.2}$$

If this condition holds, P defines a map $\xi \rightarrow P(\xi)$ of S^{n-1} into $GL(N, C)$, where S^{n-1} is the unit sphere in $\Re^n$. This is why the homotopy (and hence the homology) of $GL(N, C)$ enters into the study of elliptic operators. Now, according to the Bott periodicity theorem (Bott 1959, Atiyah and Bott 1964), the homotopy groups $\pi_{n-1}(GL(N, C))$ are 0 for n odd, and isomorphic to the integers for n even, provided that $2N \geq n$ (see also section 3.6). Thus, if n is even and $2N \geq n$, P defines an integer which may be called its *degree*. Such a degree may be evaluated explicitly as an integral

$$\int_{S^{n-1}} \omega(P),$$

where $\omega(P)$ is a differential expression in P generalizing the familiar formula $\frac{1}{2\pi i} \int_{S^1} \frac{dP}{P}$. The use of these invariants of a general elliptic operator D, jointly with the ordinary characteristic classes of M, can give an explicit formula for index(D), which is indeed similar to that occurring in the Riemann–Roch formula. As Atiyah (1966) put it, the deeper underlying reason is that, as far as the topology is concerned, the classical operators are just as complicated as the most general ones, so that the Riemann–Roch formula contains relevant information about the general case.

If one is interested in the actual proof of the index theorem (cf. section 3.1), one can say that, when M is a simple space like a sphere, Bott's theorem can be used to deform D into a standard operator, whose index may be computed directly. Moreover, if M is complicated, one can embed M in a sphere S, and construct an elliptic operator, say D', on S, such that

$$\text{index}(D) = \text{index}(D'). \tag{2.5.3}$$

Thus, reduction to the previous case is possible. However, even if one starts with a 'nice' operator on M (for example, if D results from a complex structure), it is not possible, in general, to get a nice operator D' on S. Thus, the consideration of operators which are neither 'nice' nor classical is indeed unavoidable. The topology of linear groups and the analysis of elliptic operators share important properties like linearity, stability under deformation, and finiteness. It was suggested in Atiyah (1966) that all this lies at the very heart of the close relation between these branches of mathematics. In the following chapters, we shall also study index problems for manifolds with boundary, and their deep link with general properties of complex powers of elliptic operators, and the asymptotic expansion of the integrated heat kernel (cf. Piazza 1991, 1993).

2.6 Pseudo-differential operators

Many recent developments in operator theory and spectral asymptotics deal with *pseudo-differential* operators. One then has a powerful way to analyze, under the same name, the familiar differential operators, their solution operators (i.e. their parametrices) in the elliptic case, integral and integro-differential operators, including, in particular, the singular integral operators. Examples are given by the Laplace operator $\triangle \equiv -\sum_{i=1}^{n} \frac{\partial^2}{\partial x_i^2}$ on $\Re^n$, the solution operator $(\triangle + 1)^{-1}$ defined via Fourier transform, the Newton potential (for $n \geq 2$)

$$Qh = C_n \int \frac{h(x)}{\mid x - y \mid^{n-2}} dx \tag{2.6.1}$$

for the equation $\triangle u = h$. Both the operators $(\triangle + 1)^{-1}$ and Q are of order 2, but the formalism is so general that one can define operators *of any real order*. An example presented in Grubb (1996) is the operator $(\triangle + 1)^{-s}$, defined in $L^2(\Re^n)$ with the help of spectral theory, which is a pseudo-differential operator of order $2s$, for any $s \in \Re$.

In general, any pseudo-differential operator (Seeley 1967, Shubin 1987) may be written in the form

$$A = \sum_{|\alpha| \leq m} A_\alpha D^\alpha, \tag{2.6.2}$$

where

$$D^\alpha \equiv \prod_{j=1}^{n} \left(-i \frac{\partial}{\partial x_j} \right)^{\alpha_j}, \tag{2.6.3}$$

and the α_j are such that $|\alpha| = \sum_{j=1}^{n} \alpha_j$. Their symbols are then a generalization of the *characteristic polynomial*

$$\sigma(A)(x,\xi) = \sum_{0}^{m} a_{m-j}(x,\xi), \tag{2.6.4}$$

where $a_{m-j} \equiv \sum_{|\alpha|=m-j} A_\alpha(x)\xi^\alpha$. The peculiar property of a pseudo-differential operator is that

$$\sigma(A) = \sum_{0}^{\infty} a_{m-s}, \tag{2.6.5}$$

where m may be *any* complex number, a_{m-s} is homogeneous of order $m-s$ in ξ, and the series for $\sigma(A)$ may be divergent. The physics-oriented reader might still feel unfamiliar with this abstract mathematical framework. Thus, we here summarize a less abstract presentation, due to Fulling (1996). The basic idea is of dealing with operators, say again A, whose action on functions, say ψ, can be expressed in the form

$$[A\psi](x) \equiv (2\pi\hbar)^{-\frac{m}{2}} \int_{\Re^m} dp\, e^{i\frac{\vec{p}\cdot\vec{x}}{\hbar}} a(\vec{x},\vec{p})\widehat{\psi}(p), \tag{2.6.6}$$

where $a(\vec{x},\vec{p})$ is the symbol of A, viewed as a function on the phase space $\Re^{2m}$. This symbol is required to satisfy the conditions specified after Eq. (2.6.5). Another way of defining the operator A is given by the equation

$$[A\psi](x) \equiv \int_{\Re^m} dy \mathcal{A}(\vec{x},\vec{y})\psi(\vec{y}), \tag{2.6.7}$$

where the *kernel*, $\mathcal{A}$, is obtained as

$$\mathcal{A}(\vec{x},\vec{y}) = h^{-m} \int_{\Re^m} dp\, e^{i\vec{p}\cdot(\vec{x}-\vec{y})} a(\vec{x},\vec{p}), \tag{2.6.8}$$

so that, by inverse Fourier transform, the symbol $a(\vec{x},\vec{p})$ reads

$$a(\vec{x},\vec{p}) = \int_{\Re^m} dy\, e^{i\vec{p}\cdot(\vec{y}-\vec{x})} \mathcal{A}(\vec{x},\vec{y}). \tag{2.6.9}$$

In the applications of the formalism, it may be helpful to use the *Weyl calculus*, where the kernel and the symbol are instead expressed by the formulae (Fulling 1996)

$$\mathcal{A}(\vec{x},\vec{y}) = h^{-m} \int_{\Re^m} dp\, e^{i\vec{p}\cdot(\vec{x}-\vec{y})} a\left(\frac{\vec{x}+\vec{y}}{2}, \vec{p}\right), \tag{2.6.10}$$

$$a(\vec{q},\vec{p}) = \int_{\Re^m} du\, e^{-i\vec{p}\cdot\vec{u}} \mathcal{A}\left(\vec{q}+\frac{\vec{u}}{2}, \vec{q}-\frac{\vec{u}}{2}\right). \tag{2.6.11}$$

When the pseudo-differential operators were first introduced, they were defined on functions on open sets or on manifolds without boundary. Over the past few years, however, the theory of pseudo-differential operators has made substantial progress in the case of manifolds with boundary. The key elements of the corresponding analysis are as follows (Grubb 1996). One can consider, for example, the boundary-value problem on a bounded open set $\Omega \subset \Re^n$ with smooth boundary $\partial\Omega$:

$$Pu = f \quad \text{in } \Omega, \tag{2.6.12}$$

$$Tu = \varphi \quad \text{at } \partial\Omega, \tag{2.6.13}$$

where P is the Laplace operator and T is a boundary operator (cf. section 5.2). Concrete examples for the action of T are given by Dirichlet boundary conditions

$$Tu = [u]_{\partial\Omega},$$

or Neumann boundary conditions

$$Tu = \Big[\nabla_N u\Big]_{\partial\Omega},$$

where ∇_N denotes the operator of covariant differentiation along the inward-pointing normal direction to the boundary (see section 5.2). The *boundary-value problem* is, by definition, the pair

$$\mathcal{A} = \begin{pmatrix} P \\ T \end{pmatrix} : C^\infty(\overline{\Omega}) \rightarrow C^\infty(\overline{\Omega}) \times C^\infty(\partial\Omega). \tag{2.6.14}$$

Further to this, one has to consider the *parametrix*

$$\mathcal{A}^{-1} = (R\ K) : C^\infty(\overline{\Omega}) \times C^\infty(\partial\Omega) \rightarrow C^\infty(\overline{\Omega}), \tag{2.6.15}$$

where, by definition, R is the operator such that (cf. (2.6.13))

$$Ru = 0 \quad \text{at} \quad \partial\Omega, \tag{2.6.16}$$

and K is the operator occurring in the homogeneous equation (cf. (2.6.12))

$$Ku = 0 \quad \text{in } \Omega. \tag{2.6.17}$$

The boundary-value problem described by the equations (2.6.12) and (2.6.13) is, in general, non-local. One can make it non-local in the boundary condition only, by taking T of the form

$$Tu = \beta_1 u + S\beta_0 u, \tag{2.6.18}$$

where S is a first-order pseudo-differential operator on the boundary. One can also make one of the two choices (Grubb 1996)

$$Tu = \beta_0 u + T_0' u, \tag{2.6.19}$$

$$Tu = \beta_1 u + T_1' u, \tag{2.6.20}$$

where T_0' and T_1' are *integral operators* mapping Ω into $\partial\Omega$, and β_0 and β_1 are yet other operators of suitable order. The boundary operators (2.6.19) and (2.6.20) lead to greater technical difficulties with respect to the operator (2.6.18).

An enlightening example of non-local boundary conditions is provided by a problem studied by Schröder (1989). The basic properties are as follows. Given, in $L_1(\Re) \cap L_2(\Re)$, the function q, one defines

$$q_R(x) \equiv \frac{1}{2\pi R} \sum_{l=-\infty}^{\infty} e^{ilx/R} \int_{-\infty}^{\infty} e^{-ily/R} q(y) dy. \tag{2.6.21}$$

By construction, q_R is periodic with period $2\pi R$, and tends to q as $R \to \infty$. One then studies the Laplace operator Δ in $L_2(B_R)$, B_R being the set defined by

$$B_R \equiv \left\{ x, y : x^2 + y^2 \leq R^2 \right\}, \tag{2.6.22}$$

with non-local boundary conditions

$$\left[\frac{\partial u}{\partial n} \right]_{\partial B_R} + \oint_{\partial B_R} q_R(s - s') u(R \cos(s'/R), R \sin(s'/R)) ds' = 0, \tag{2.6.23}$$

where

$$x \equiv R \cos(s/R), \tag{2.6.24}$$

$$y \equiv R \sin(s/R). \tag{2.6.25}$$

The self-adjointness of Δ requires that $q(s) = \overline{q(-s)}$. To solve the eigenvalue equation $\Delta u = Eu$, polar coordinates are used, so that the boundary-value problem reads

$$-\left(\frac{\partial^2 u}{\partial r^2} + \frac{1}{r} \frac{\partial u}{\partial r} + \frac{1}{r^2} \frac{\partial^2 u}{\partial \varphi^2} \right) = Eu, \tag{2.6.26}$$

$$\frac{\partial u}{\partial r}(R, \varphi) + R \int_{-\pi}^{\pi} q_R(R(\varphi - \theta)) u(R, \theta) d\theta = 0. \tag{2.6.27}$$

The spectrum of Δ consists of a positive ($E > 0$) and a negative

($E < 0$) part. In the former case, the eigenfunctions take the form (J_l denoting, as usual, the Bessel functions of order l)

$$u_{l,E}(r,\varphi) = J_l(r\sqrt{E})e^{il\varphi}, \; l \in Z, \tag{2.6.28}$$

up to a multiplicative constant. The insertion of (2.6.28) into (2.6.27) leads to

$$\begin{aligned} 0 &= \sqrt{E}J_l'(R\sqrt{E})e^{il\varphi} + RJ_l(R\sqrt{E})\int_{-\pi}^{\pi} q_R(R(\varphi-\theta))e^{il\theta}d\theta \\ &= \left(\sqrt{E}J_l'(R\sqrt{E}) + RJ_l(R\sqrt{E})\int_{-\pi}^{\pi} q_R(R\alpha)e^{-il\alpha}d\alpha\right)e^{il\varphi} \\ &= \left(\sqrt{E}J_l'(R\sqrt{E}) + J_l(R\sqrt{E})\tilde{q}(l/R)\right)e^{il\varphi}, \end{aligned} \tag{2.6.29}$$

where we have used the definition (2.6.21), and $\tilde{q}$ denotes the Fourier transform of q. Hence the eigenvalues obey the equation

$$\sqrt{E}J_l'(R\sqrt{E}) + J_l(R\sqrt{E})\tilde{q}(l/R) = 0. \tag{2.6.30}$$

It is now convenient to define

$$w \equiv R\sqrt{E}, \tag{2.6.31}$$

so that Eq. (2.6.30) reads

$$F_l(w) \equiv w\frac{J_l'(w)}{J_l(w)} = -R\tilde{q}(l/R). \tag{2.6.32}$$

Note now that the function F_l has single poles at the points $w = j_{l,k}$, i.e. the zeroes of J_l, is elsewhere smooth, and has negative first derivative. This implies that there exists a sequence $w_{l,k}(R)$ of solutions of Eq. (2.6.32) such that

$$w_{l,k}(R) \in]j_{l,k}', j_{l,k+1}'[, \tag{2.6.33}$$

where $j_{l,k}'$ are the zeroes of J_l'. On the other hand, since

$$J_l(w) \sim \frac{(w/2)^l}{\Gamma(l+1)} \text{ as } w \to 0,$$

one finds

$$\lim_{w\to 0} F_l(w) = \mid l \mid . \tag{2.6.34}$$

Thus, if the following inequality holds:

$$-\frac{\mid l \mid}{R} \leq \tilde{q}(l/R) < 0, \tag{2.6.35}$$

then an additional solution exists, say $w_l(R)$, such that

$$w_l(R) \in [0, j_{l,1}'[. \tag{2.6.36}$$

Correspondingly, one finds eigenvalues $E_l(R) = w_l^2(R)/R^2$ (see (2.6.31)) whose eigenfunctions decrease exponentially with increasing distance d from the boundary (at least if $d << R$). They are called *surface states.* By contrast, the eigenvalues $E_{l,k}(R) = w_{l,k}^2(R)/R^2$ resulting from (2.6.33) correspond to more extended eigenfunctions, and are called *bulk states.*

For negative values of the energy, the eigenfunctions read instead (cf. (2.6.28))

$$u_{l,E}(r,\varphi) = I_l(r\sqrt{-E})e^{il\varphi}, \; l \in Z, \tag{2.6.37}$$

where I_l is the standard notation for modified Bessel functions of first kind of order l. The resulting eigenvalue condition is now

$$G_l(z) \equiv z\frac{I_l'(z)}{I_l(z)} = -R\tilde{q}(l/R), \tag{2.6.38}$$

where $z \equiv R\sqrt{-E}$. The relevant asymptotic behaviours are

$$\lim_{z \to 0} G_l(z) = \mid l \mid, \tag{2.6.39}$$

$$G_l(z) \sim (l^2 + z^2)^{1/2} \text{ as } l \to \infty. \tag{2.6.40}$$

Bearing in mind that $G_l'(z) > 0$ if $z > 0$, one finds that Eq. (2.6.38) can have only one solution, provided that

$$\tilde{q}(l/R) < -\frac{\mid l \mid}{R}. \tag{2.6.41}$$

Further details on surface states, bulk states and their relevance for physical applications, can be found in the paper by Schröder (1989).

3
Index problems

This chapter begins with an outline of index problems for closed manifolds and for manifolds with boundary, and of the relation between index theory and anomalies in quantum field theory. In particular, the index of the Dirac operator is evaluated in the presence of gauge fields and gravitational fields. The result is expressed by integrating the Chern character and the Dirac genus. Interestingly, anomaly calculations turn out to be equivalent to the analysis of partition functions in ordinary quantum mechanics. The chapter ends with an introductory presentation of Bott periodicity and K-theory, with the aim of describing why the Dirac operator is the most fundamental, in the theory of elliptic operators on compact Riemannian manifolds.

3.1 Index problem for manifolds with boundary

Following Atiyah and Bott (1965), this section is devoted to some aspects of the extension of the index formula for closed manifolds, to manifolds with boundary. The naturally occurring question is then how to measure the topological implications of elliptic boundary conditions, since boundary conditions have of course a definite effect on the index.

For example, let $\mathcal{Q}$ be the unit disk in the plane, let Y be the boundary of $\mathcal{Q}$, and let b be a nowhere-vanishing vector field on Y. Denoting by D the Laplacian on $\mathcal{Q}$, we consider the operator

$$(D,b) : C^\infty(\mathcal{Q}) \rightarrow C^\infty(\mathcal{Q}) \oplus C^\infty(Y). \tag{3.1.1}$$

By definition, (D,b) sends f into $Df \oplus (bf \mid Y)$, where bf is the directional derivative of f along b. Since D is elliptic and b is non-vanishing, kernel and cokernel of (D,b) are both finite-dimensional, hence the index of the boundary-value problem is finite. One would now like it to express the index in terms of the topological data given by D and the boundary conditions. The solution of this problem is expressed by a formula derived by Vekua (Hörmander 1963), according to which

$$\text{index}\,(D,b) = 2\Big(1 - \text{winding number of } b\Big). \tag{3.1.2}$$

It is now necessary to recall the index formula for closed manifolds, and this is here done in the case of vector bundles. From now on, M is the base manifold, Y its boundary, $T(M)$ the tangent bundle of M, $B = B(M)$ the ball-bundle (consisting of vectors in $T(M)$ of length ≤ 1), $S = S(M)$ the sphere-bundle of unit vectors in $T(M)$. Of course, if M is a closed manifold (hence without boundary), then $S(M)$ is just the boundary of $B(M)$, i.e.

$$\partial B(M) = S(M) \quad \text{if } \partial M = \emptyset. \tag{3.1.3}$$

However, if M has a boundary Y, then

$$\partial B(M) = S(M) \cup B(M) \mid Y, \tag{3.1.4}$$

where $B(M) \mid Y$ is the subspace of $B(M)$ lying over Y under the natural map $\pi : B(M) \rightarrow M$.

We now focus on a system of k linear partial differential operators, i.e.

$$Df_i \equiv \sum_{j=1}^{k} A_{ij} f_j, \tag{3.1.5}$$

defined on the n-dimensional manifold M. As we know from section 2.3, the leading symbol of D is a function $\sigma(D)$ on $T^*(M)$ attaching to each cotangent vector λ of M, the matrix $\sigma(D:\lambda)$ obtained from the highest terms of A_{ij} by replacing $\frac{\partial^\alpha}{\partial x^\alpha}$ with $(i\lambda)^\alpha$. Moreover, the system D is *elliptic* if and only if the function $\sigma(D)$ maps $S(M)$ into the group $GL(k,C)$ of non-singular $k \times k$ matrices with complex coefficients. Thus, defining

$$GL \equiv \lim_{m \rightarrow \infty} GL(m,C), \tag{3.1.6}$$

the leading symbol defines a map

$$\sigma(D) : S(M) \rightarrow GL. \tag{3.1.7}$$

Since the index of an elliptic system is invariant under deformations, on a closed manifold the index of D only depends on the homotopy class of the map $\sigma(D)$.

To write down the index formula we are looking for, one constructs a definite differential form $ch \equiv \sum_i ch^i$ on GL. Note that, strictly, we define a differential form $ch(m)$ on each $GL(m, C)$ (see Eqs. (3.B.12) and (3.B.13) for the Chern character). Moreover, by using a universal expression in the curvature of M, one constructs the Todd class of $T(M)$, denoted by $td(M)$, as in (3.B.16). The index theorem for a closed manifold then yields the index of D as an integral

$$\text{index}(D) = \int_{S(M)} \sigma(D)^* \, ch \wedge \pi^* td(M). \tag{3.1.8}$$

With our notation, $\sigma(D)^* ch$ is the form on GL pulled back to $S(M)$ via $\sigma(D)$. By virtue of (3.1.3), the integral (3.1.8) may be re-written as

$$\text{index}(D) = \int_{\partial B(M)} \sigma(D)^* \, ch \wedge \pi^* td(M). \tag{3.1.9}$$

In this form the index formula is also meaningful for a manifold with boundary, *provided that* $\sigma(D)$, originally defined on $S(M)$, is *extended* to $\partial B(M)$. It is indeed in this extension that the topological data of a set of elliptic boundary conditions manifest themselves. Following Atiyah and Bott (1965), we may in fact state the following theorem (cf. Booss and Bleecker 1985).

Theorem 3.1.1 A set of elliptic boundary conditions, B, on the elliptic system D, defines a definite map $\sigma(D, B) : \partial B(M) \rightarrow GL$, which extends the map $\sigma(D) : S(M) \rightarrow GL$ to the whole of $\partial B(M)$.

One thus finds an index theorem formally analogous to the original one, in that the index of D subject to the elliptic boundary condition B is given by

$$\text{index}(D, B) = \int_{\partial B(M)} \sigma(D, B)^* ch \wedge \pi^* td(M). \tag{3.1.10}$$

One can thus appreciate a key feature of index theorems: they relate analytic properties of differential operators on fibre bundles to the topological invariants of such bundles (i.e. their characteristic classes). As will be shown in the following sections, the relevance of index theorems for theoretical physics lies in the possibility to evaluate the number of zero-modes of differential operators, if one knows the topology of the fibre bundles under consideration.

3.2 Elliptic boundary conditions

Although this section is (a bit) technical, it is necessary to include it for the sake of completeness. Relevant examples of elliptic boundary-value problems (see appendix 3.A) will be given in chapters 4–6.

Following Atiyah and Bott (1965), let D be a $k \times k$ elliptic system of differential operators on M, and let r denote the order of D. The symbol $\sigma(D)$ is a function on the cotangent vector bundle $T^*(M)$ whose values are $(k \times k)$-matrices, and its restriction to the unit sphere-bundle $S(M)$ takes non-singular values. For our purposes, it is now necessary to consider a system B of boundary operators (cf. (5.1.18)) given by an $l \times k$ matrix with rows $b_1, ..., b_l$ of orders $r_1, ..., r_l$ respectively. We now denote by $\sigma(b_i)$ the symbol of b_i, and by $\sigma(B)$ the matrix with $\sigma(b_i)$ as ith row. At a point of the boundary Y of M, let ν be the unit inward-pointing normal and let y denote any unit tangent vector to Y. One puts

$$\sigma_y(D)(t) = \sigma(D)(y + t\nu), \tag{3.2.1}$$

$$\sigma_y(B)(t) = \sigma(B)(y + t\nu), \tag{3.2.2}$$

so that $\sigma_y(D)$ and $\sigma_y(B)$ are polynomials in t. It then makes sense to consider the system of ordinary linear equations

$$\sigma_y(D)\left(- i\frac{d}{dt}\right)u = 0, \tag{3.2.3}$$

whose space of solutions is denoted by $\mathcal{M}_y$. The *ellipticity* of the system D means, by definition, that $\mathcal{M}_y$ can be decomposed as

$$\mathcal{M}_y = \mathcal{M}_y^+ \oplus \mathcal{M}_y^-, \tag{3.2.4}$$

where $\mathcal{M}_y^+$ consists only of exponential polynomials involving $e^{i\lambda t}$ with $\text{Im}(\lambda) > 0$, and $\mathcal{M}_y^-$ involves those with $\text{Im}(\lambda) < 0$. The ellipticity condition for the system of boundary operators, relative to the elliptic system of differential operators, is that the equations (3.2.3) should have a *unique* solution $u \in \mathcal{M}_y^+$ satisfying the boundary condition (cf. appendix 3.A)

$$\sigma_y(B)\left(- i\frac{d}{dt}\right)u\,|_{t=0} = V, \tag{3.2.5}$$

for any given $V \in C^l$.

Now a lemma can be proved, according to which a natural isomorphism of vector spaces exists between M_y^+, the cokernel of $\sigma_y(D)(t)$, and $\mathcal{M}_y^+$:

$$M_y^+ \cong \mathcal{M}_y^+. \tag{3.2.6}$$

Hence the elliptic boundary condition yields an isomorphism

$$\beta_y^+ : M_y^+ \to C^l. \tag{3.2.7}$$

The set of all M_y^+, for $y \in S(Y)$, forms a vector bundle M^+ over $S(Y)$, and (3.2.7) defines an isomorphism β^+ of M^+ with the trivial bundle $S(Y) \times C^l$.

As we know from section 2.6, possible boundary conditions are differential or integro-differential. If only differential boundary conditions are imposed then the map β_y^+, regarded as a function of y, cannot be an arbitrary continuous function. To obtain *all* continuous functions one has to enlarge the problem and consider integro-differential boundary conditions as well. This is, indeed, an important topological simplification. Thus, an elliptic problem (D, B) has associated with it $\sigma(D), M^+, \beta^+$, where β^+ can be any vector-bundle isomorphism of M^+ with the trivial bundle.

3.3 Index theorems and gauge fields

Anomalies in gauge theories, and their relation to the index theory of the Dirac operator, are a key topic in the quantization of gauge theories of fundamental interactions. The aim of this section is to introduce the reader to ideas and problems in this field of research, following Atiyah (1984b) and then Alvarez-Gaumé (1983b).

The index theorem is concerned with any elliptic differential operator, but for physical applications one only needs the special case of the Dirac operator. From chapter 1, we know that this is globally defined on any Riemannian manifold (M, g) provided that M is oriented and has a spin-structure. If M is compact, the Dirac operator is elliptic, self-adjoint, and has a discrete spectrum of eigenvalues. Of particular interest is the 0-eigenspace (or null-space). By definition, the Dirac operator acts on spinor fields, and it should be emphasized that there is a basic algebraic difference between spinors in even and odd dimensions. In fact in odd dimensions spinors belong to an irreducible representation of the spin-group, whereas in even dimensions spinors break up into two irreducible pieces, here denoted by S^+ and S^-. The Dirac operator interchanges these two, hence it consists essentially of an operator (cf. section 1.4)

$$D : S^+ \rightarrow S^-, \tag{3.3.1}$$

and its adjoint

$$D^* : S^- \rightarrow S^+. \tag{3.3.2}$$

The dimensions of the null-spaces of D and D^* are denoted by N^+ and N^- respectively, and the index of D (cf. Eqs. (2.3.3)–(2.3.5)) is defined by

$$\text{index}(D) \equiv N^+ - N^-. \tag{3.3.3}$$

Note that, although a formal symmetry exists between positive and negative spinors (in fact they are interchanged by reversing the orientation of M), the numbers N^+ and N^- need not be equal, due to topological effects. Hence the origin of the *chiral anomaly* (cf. Rennie 1990).

The index theorem provides an explicit formula for index(D) in terms of topological invariants of M. Moreover, by using the Riemannian metric g and its curvature tensor $R(g)$, one can write an explicit integral formula

$$\text{index}(D) = \int_M \Omega(R(g)), \tag{3.3.4}$$

where Ω is a formal expression obtained purely algebraically from $R(g)$. For example, when M is four-dimensional one finds

$$\Omega = \frac{\text{Tr}\,\omega^2}{96\pi^2}, \tag{3.3.5}$$

where ω is regarded as a matrix of two-forms. Interestingly, the index formula (3.3.4) is, from the physical point of view, purely gravitational since it only involves the metric g. Moreover, the index vanishes for the sphere or torus, and one needs more complicated manifolds to exhibit a non-vanishing index. However, (3.3.4) can be generalized to gauge theories. This means that one is given a complex vector bundle V over M with a unitary connection A. If the fibre of V is isomorphic with C^N, then A is a

$U(N)$ gauge field over M. The *extended Dirac operator* now acts on vector-bundle-valued spinors

$$D_A : S^+ \otimes V \to S^- \otimes V, \tag{3.3.6}$$

and one defines again the index of D_A by a formula like (3.3.3). The index formula (3.3.4) is then replaced by

$$\text{index}\Big(D_A\Big) = \int_M \Omega(R(g), F(A)), \tag{3.3.7}$$

where $F(A)$ is the curvature of A, and $\Omega(R, F)$ is a certain algebraic expression in R and F. For example, if the manifold M is four-dimensional, (3.3.5) is generalized by

$$\Omega = \frac{\text{Tr}\,\omega^2}{96\pi^2} - \frac{\text{Tr}\,F^2}{8\pi^2}. \tag{3.3.8}$$

We now find it instructive, instead of giving a brief outline of a wide range of results (see, for example, Rennie 1990 and Gilkey 1995), to focus on a specific but highly non-trivial example. Following Alvarez-Gaumé (1983b), we deal (again) with the Dirac equation in the presence of external gauge and gravitational fields. The motivations of our investigation lie in the deep link between supersymmetry and the Atiyah–Singer index theorem. Indeed, if one considers supersymmetric quantum mechanics, which may be viewed as a (0+1)-dimensional field theory, one finds that this theory has N conserved charges, say Q_i, which anticommute with the fermion number operator $(-1)^F$, and which satisfy the supersymmetry algebra (for all $i, j = 1, ..., N$):

$$\Big\{Q_i, Q_j^*\Big\} = 2\delta_{ij} H, \tag{3.3.9}$$

$$\Big\{Q_i, (-1)^F\Big\} = 0, \tag{3.3.10}$$

$$\{Q_i, Q_j\} = 0, \tag{3.3.11}$$

where H is the Hamiltonian of supersymmetric quantum mechanics. If Q, Q^* are any of the N supersymmetric charges, the operator

$$S \equiv \frac{1}{\sqrt{2}}(Q + Q^*) \tag{3.3.12}$$

is Hermitian and satisfies $S^2 = H$. Moreover, if φ_E is an arbitrary eigenstate of H:

$$H\varphi_E = E\varphi_E, \tag{3.3.13}$$

then $S\varphi_E$ is another state with the same energy. Thus, if φ_E is a bosonic state, $S\varphi_E$ is fermionic, and the other way around, and non-zero energy states in the spectrum appear in Fermi–Bose pairs. This in turn implies that $\mathrm{Tr}(-1)^F e^{-\beta H}$ receives contributions only from zero-energy states. Further to this, the work in Witten (1982) has shown that $\mathrm{Tr}(-1)^F e^{-\beta H}$ is a *topological invariant* of the quantum theory. Bearing in mind that bosonic zero-energy states, say φ_B, are, by definition, solutions of the equation

$$Q\varphi_B = 0, \tag{3.3.14}$$

and fermionic zero-modes solve instead the equation

$$Q^* \varphi_F = 0, \tag{3.3.15}$$

one finds that

$$\mathrm{Index}(Q) = \dim \mathrm{Ker}\, Q - \dim \mathrm{Ker}\, Q^* = \mathrm{Tr}\left[(-1)^F e^{-\beta H}\right]. \tag{3.3.16}$$

The problem of finding the index of Q is therefore reduced to the evaluation of the trace on the right-hand side of Eq. (3.3.16) in the $\beta \rightarrow 0$ limit. The idea of Alvarez-Gaumé (1983a,b) was to perform this calculation with the help of a path-integral representation of the partition function, with periodic boundary conditions for fermionic fields:

$$\mathrm{Tr}\left[(-1)^F e^{-\beta H}\right] = \int_{\Omega_\beta} d\phi(t)\, d\psi(t) \exp\left(-\int_0^\beta L_E(t) dt\right), \tag{3.3.17}$$

where ϕ and ψ are required to belong to the class Ω_β of fermionic fields, say χ, such that

$$\chi(0) = \chi(\beta). \tag{3.3.18}$$

The next non-trivial step is to use the Lagrangian of the supersymmetric non-linear σ-model obtained by dimensional reduction from $1+1$ to $0+1$ dimensions:

$$L = \frac{1}{2} g_{ij}(\phi) \dot{\phi}^i \dot{\phi}^j + \frac{i}{2} g_{jk} \psi_\alpha^j \left[\frac{d}{dt}\psi_\alpha^k + \Gamma^k{}_{ml}\, \dot{\phi}^m\, \psi_\alpha^l\right]$$
$$+ \frac{1}{4} R_{ijkl}\, \psi_1^i\, \psi_1^j\, \psi_2^k\, \psi_2^l. \tag{3.3.19}$$

With this notation, g_{ij} is the metric on the manifold M, $\Gamma^i{}_{jk}$ are the Christoffel symbols, ${R^i}_{jkl}$ is the Riemann tensor of M, $\psi_\alpha^i(t)$ are real anticommuting fermionic fields ($\alpha = 1, 2$).

Since our ultimate goal is the analysis of the Dirac equation in the presence of gauge fields, it is now appropriate to consider a gauge group, say G, acting on M, with gauge connection $A_i^\alpha(\phi)$ and gauge curvature

$$F_{ij}^\alpha(\phi) = \partial_i A_j^\alpha - \partial_j A_i^\alpha + h f_{\beta\gamma}^\alpha \, A_i^\beta \, A_j^\gamma, \tag{3.3.20}$$

where h is the gauge coupling constant and Greek indices range from 1 through $\dim(G)$. In the explicit expression of the first-order derivative resulting from the Dirac operator, one has now to include both the effects of A_i^α and the contribution of the spin-connection, say $\omega_{\ ib}^a$. In terms of the vierbein $e_{\ i}^a$ (cf. section 1.2), and of its dual ${E^i}_a$:

$$E^i_{\ a} \, e^a_{\ j} = \delta^i_{\ j}, \tag{3.3.21}$$

one has

$$\omega^a_{\ ib} = -E_b^k \Big(\partial_i e^a_{\ k} - {\Gamma^l}_{ik} \, e^a_{\ l} \Big), \tag{3.3.22}$$

and the eigenvalue problem for the Dirac operator reads

$$i\gamma^k \Big(\partial_k + \frac{1}{2} \omega_{kab} \sigma^{ab} + ih A_k^\alpha T^\alpha \Big)_{AB} \Big(\psi_\lambda \Big)_B = \lambda \Big(\psi_\lambda \Big)_A. \tag{3.3.23}$$

In Eq. (3.3.23), one has (cf. (1.2.10)) $\sigma^{ab} \equiv \frac{1}{4}\Big[\gamma^a, \gamma^b\Big]$, and $\Big(T^\alpha\Big)_{AB}$ are the generators of the representation of G, with indices A, B ranging from 1 through $\dim(T)$. The reader should be aware that our a, b indices, here used to follow the notation in Alvarez-Gaumé (1983b), correspond to the hatted indices of section 1.2.

We are eventually interested in the one-dimensional analogue of Eq. (3.3.23). For this purpose, we consider a pair of fermionic annihilation and creation operators, say C_A, C_A^*:

$$\{C_A, C_B\} = \{C_A^*, C_B^*\} = 0, \tag{3.3.24}$$

$$\{C_A^*, C_B\} = \delta_{AB}. \tag{3.3.25}$$

They lead to the eigenvalue equation

$$i\gamma^k \Big(\partial_k + \frac{1}{2} \omega_{kab} \sigma^{ab} + ih A_k^\alpha C^* T C \Big) \psi_\lambda = \lambda \psi_\lambda. \tag{3.3.26}$$

Interestingly, if ψ_λ is an eigenfunction of the Dirac operator with eigenvalue $\lambda \neq 0$, then $\gamma_5 \psi_\lambda$ turns out to be the eigenfunction belonging to the eigenvalue $-\lambda$. Thus, since (hereafter, $\mathcal{D} \equiv i\gamma^k D_k$) $[\mathcal{D}^2, \gamma_5] = 0$, one finds (see (3.3.16))

$$\mathrm{Tr}\left[\gamma_5 e^{-\beta \mathcal{D}^2}\right] = n_{E=0}(\gamma_5 = 1) - n_{E=0}(\gamma_5 = -1), \qquad (3.3.27)$$

where $n_{E=0}(\gamma_5 = 1)$ and $n_{E=0}(\gamma_5 = -1)$ denote the number of zero-eigenvalues of $\mathcal{D}$ with $\gamma_5 = 1$ and $\gamma_5 = -1$, respectively. On writing this equation, we have recognized that $(-1)^F = \gamma_5$ in four dimensions. Now the Euclidean Lagrangian is inserted into the path-integral formula (3.3.17), with the understanding that functional integration is restricted to the space of one-particle states for the (C, C^*) fermions. This restriction is indeed necessary to ensure that the index of $\mathcal{D}$ is evaluated only in the representation T^α of the gauge group G. The boundary conditions are now (3.3.18) for ϕ and ψ, jointly with antiperiodic boundary conditions for fermionic operators:

$$C(0) = -C(\beta), \; C^*(0) = -C^*(\beta). \qquad (3.3.28)$$

Moreover, the background-field method is used, writing (recall that $\psi^a \equiv e^a_{\ i} \, \psi^i$)

$$\phi^i = \phi^i_0 + \xi^i, \; \psi^a = \psi^a_0 + \eta^a. \qquad (3.3.29)$$

The underlying idea is that, in the $\beta \rightarrow 0$ limit, the functional integral is dominated by time-independent field configurations ϕ^i_0 and ψ^a_0. The terms ξ^i and η^a are then the fluctuations around the constant background values, according to (3.3.29). Omitting a few details, which can be found in Alvarez-Gaumé (1983b) and references therein, one obtains

$$\mathrm{index}(\mathcal{D}) = (i/2\pi)^n \int \mathrm{dvol} \int d\psi_0 \left(\mathrm{Tr} \exp\Bigl(-\frac{i}{2} \psi^a_0 \psi^b_0 F^\alpha_{\ ab} T^\alpha \Bigr) \right)$$
$$\times \prod_{l=1}^{n} \frac{(ix_l/2)}{\sinh(ix_l/2)}, \qquad (3.3.30)$$

where $(2\pi)^{-n}$ results from the Feynman measure for constant modes, $(i)^n$ is due to integration over constant and real-valued fermionic configurations, and the x_l are the eigenvalues of the matrix

$$\frac{1}{2} R_{abcd} \, \psi^c_0 \, \psi^d_0 .$$

Note that, considering the Riemann curvature two-form

$$R_{ab} \equiv \frac{1}{2} R_{abcd} \, e^c \wedge e^d, \tag{3.3.31}$$

and the gauge curvature two-form

$$F \equiv \frac{h}{2} F^{\alpha}_{ab} \, T^{\alpha} \, e^a \wedge e^b, \tag{3.3.32}$$

one can form the polynomials (cf. appendix 3.B)

$$ch(F) \equiv \text{Tr } e^{F/2\pi}, \tag{3.3.33}$$

$$\widehat{A}(M) \equiv \prod_{\alpha} \frac{(\omega_\alpha/4\pi)}{\sinh(\omega_\alpha/4\pi)}. \tag{3.3.34}$$

In the definition (3.3.33), the trace is taken over the representation T^α of G, and in (3.3.34) the ω_α are the eigenvalues of the antisymmetric matrix (3.3.31). According to a standard notation, $ch(F)$ is the Chern character of the principal bundle associated to the gauge field, and $\widehat{A}(M)$ is the Dirac genus of M. They make it possible to re-express the result (3.3.30) in the neat form (cf. (3.1.8))

$$\text{index}(\mathcal{D}) = \int_M ch(F) \widehat{A}(M). \tag{3.3.35}$$

In particular, if M is four-dimensional, Eq. (3.3.35) leads to (cf. (3.3.7))

$$\text{index}(\mathcal{D}) = \frac{(\text{dim}T)}{192\pi^2} \int_M \text{Tr} R \wedge R + \frac{1}{8\pi^2} \int_M \text{Tr} F \wedge F. \tag{3.3.36}$$

As Alvarez-Gaumé (1983b) pointed out, this calculation has great relevance for the theory of anomalies in quantum field theory. Indeed, anomaly calculations involve usually traces of the form $\sum_n \psi_n^\dagger(x) \, L \, \psi_n(x)$, where L is a differential or algebraic operator, and the $\psi_n' s$ are eigenfunctions of the Dirac operator in the presence of external fields (either gravitational or gauge fields). The analysis leading to (3.3.36) has shown that such traces can be turned into functional integrals for one-dimensional field theories. In other words, anomaly calculations are equivalent to the analysis of partition functions in quantum mechanics (Alvarez-Gaumé 1983b).

Indeed, another definition of index is also available in the literature. This is motivated by the analysis of a family D of elliptic operators D_x, parametrized by the points x of a compact space, say M. The corresponding index is then defined by (Rennie 1990)

$$\text{index}(D) \equiv \text{Ker}(D) - \text{Coker}(D), \tag{3.3.37}$$

where Ker(D) denotes the *family* of vector spaces Ker(D_x), and similarly for Coker(D). The index defined in (3.3.37) turns out to be an element of the K-theory of M (see section 3.7). An outstanding open problem is to prove that the supersymmetric functional integral admits a K-theoretic interpretation. Moreover, it remains to be seen whether supersymmetry can be applied to prove the index theorem for families (cf. Atiyah and Singer 1971a, Bismut 1986a,b). The author is indebted to Professor L. Alvarez-Gaumé for correspondence about these issues.

3.4 Index of two-parameter families

A very important property of the index is that *it is unchanged by perturbation of the operator*, hence it can be given by a topological formula. This is why, when we vary the gauge field A continuously, the index of the extended Dirac operator D_A does not change. It is, however, possible to extract more topological information from a continuous family of operators, and we here focus on the two-parameter case (Atiyah 1984b).

For this purpose we consider a two-dimensional surface Σ compact, connected and oriented, and suppose a fibre bundle Y over Σ is given with fibre F. The twist involved in making a non-trivial bundle is an essential topological feature. By giving Y a Riemannian metric, one gets metrics g_x on all fibres F_x, hence one has metrics on F parametrized by Σ. The corresponding family of Dirac operators on F is denoted by D_x. We can now define two topological invariants, i.e. index(D_x) and the *degree* of the family of Dirac operators. To define such a degree, we restrict ourselves to the particular case when the index vanishes.

Since index$(D_x) = 0$ by hypothesis, the generic situation (one can achieve this by perturbing the metric on Y) is that D_x is invertible $\forall x \in \Sigma$ except at a finite number of points $x_1, ..., x_k$. Moreover, a multiplicity ν_i can be assigned to each x_i, which is generically ± 1. To obtain this one can, in the neighbourhood of each x_i, replace D_x by a finite matrix T_x. Zeros of det(T_x) have multiplicity ν_i at x_i. We are actually dealing with the phase-variation of $\det(T_x)$ as x traverses positively a small circle centred at x_i.

The degree ν of the family of Dirac operators D_x is defined by

$$\nu \equiv \sum_{i=1}^{k} \nu_i. \qquad (3.4.1)$$

This is a topological invariant of the family in that it is invariant under perturbation. Hence it only depends on the topology of the fibre bundle Y and vanishes for the product $F \times \Sigma$. In the case of gauge fields, one can fix the metric g on Σ and take $Y = F \times \Sigma$, but we also take a vector bundle V on Y with a connection. This gives a family A_x of gauge fields, and the corresponding Dirac family has again a degree which now depends on the topology of V. Such a degree vanishes if the vector bundle V comes from a bundle on F.

3.5 Determinants of Dirac operators

For fermionic fields one has to define determinants of Dirac operators, and these are not positive-definite by virtue of their first-order nature. Consider an even-dimensional Riemannian manifold (M,g) with Dirac operators $D_g : S^+ \rightarrow S^-$, depending on the metric g. We look for a regularized complex-valued determinant $\det(D_g)$ which should have the following properties (Atiyah 1984b):

(1) $\det(D_g)$ is a differentiable function of g;

(2) $\det(D_g)$ is gauge-invariant, i.e. $\det\left(D_{f(g)}\right) = \det(D_g)$ for any diffeomorphism f of M;

(3) $\det(D_g) = 0$ to first-order if and only if $D_g^* \, D_g$ has exactly one zero-eigenvalue.

The third property is the characteristic property of determinants in finite dimensions and it is a *minimum requirement* for any regularized determinant. Note also that, since $D_g^* \, D_g$ is an operator of Laplace type, there is no difficulty in defining its determinant, so that one can also define $|\det(D_g)|$. The problem is to define the *phase* of $\det(D_g)$ (cf. section 6 of Martellini and Reina 1985).

Suppose now that one can find a fibration over a surface Σ with fibre F, so that the Dirac operators D_x all have vanishing index whereas the *degree* of the family does not vanish. This implies that there is no function $\det(D_g)$ with the three properties just listed. Suppose in fact that such a $\det(D_g)$ exists. Then (1) and (2) imply that $h(x) \equiv \det\left(D_{g_x}\right)$ is a differentiable complex-valued function on Σ. Moreover, (3) implies that ν is the number of zeros of h counted with multiplicities and signs, i.e. that ν is the degree of $h : \Sigma \rightarrow C$. But topological arguments may be used to show that the degree of h has to vanish, hence $\nu = 0$, contradicting the assumption.

We should now bear in mind that the local multiplicities ν_i at points x_i where D_x is not invertible, are defined as local phase variations obtained by using an eigenvalue cut-off. Remarkably, if it were possible to use such a cut-off consistently at all points, the total phase variation ν would have to vanish. Thus, a fibration with non-vanishing degree implies a behaviour of the eigenvalues which prevents a good cut-off regularization. Such a remark is not of purely academic interest, since two-dimensional examples can be found, involving surfaces F and Σ of fairly high genus, where all these things happen. What is essentially involved is the variation in the conformal structure of Σ (cf. section 5.3).

Other examples of non-vanishing degree, relevant for gauge theory, are obtained when F is the $2n$-sphere S^{2n}, $\Sigma = S^2$ and the vector bundle V is $(n+1)$-dimensional, so that the gauge group G is $U(n+1)$ or $SU(n+1)$. If one represents S^2 as $\Re^2 \cup \infty$ then, removing the point at infinity, one obtains a two-parameter family of gauge fields on F parametrized by $\Re^2$. Since this family is defined by a bundle over $F \times S^2$, gauge fields are all gauge-equivalent as one goes to ∞ in $\Re^2$. What are we really up to? By taking a large circle in $\Re^2$, one gets a closed path in the group $\mathcal{G}$ of gauge transformations of the bundle on F. The degree ν is essentially a homomorphism $\pi_1(\mathcal{G}) \rightarrow Z$, and if ν does not vanish, the variation in phase of $\det(D_g)$ going round this path in $\mathcal{G}$ proves the lack of gauge-invariance. If $F = S^{2n}$ one has

$$\pi_1(\mathcal{G}) = \pi_{2n+1}(G), \tag{3.5.1}$$

$$\pi_{2n+1}\Big(SU(n+1)\Big) = Z, \tag{3.5.2}$$

and the isomorphism is given by ν, by virtue of the general index theorem for families (Atiyah and Singer 1971a).

3.6 Bott periodicity

A crucial and naturally occurring problem in all our investigations is why the Dirac operator is so important in index theory. The answer is provided by the so-called Bott periodicity theorem. We now find it helpful, for the general reader, to devote some efforts to a detailed description of how one arrives at the Bott periodicity theorem. For this purpose, we shall rely on the presentation in Milnor (1963a) (cf. Rennie 1990). Indeed, there exist two periodicity theorems: the former holds for the unitary group, while the latter is proved for the orthogonal group. We first focus on the former, and we begin with an elementary review of properties of the exponential map.

For any $n \times n$ complex matrix, say A, its exponential is defined by the convergent power series expansion (here $A^0 \equiv I$)

$$e^A = \sum_{n=0}^{\infty} \frac{A^n}{n!}. \tag{3.6.1}$$

In particular, we are interested in the following elementary properties (the dagger denoting complex conjugation and transposition)

(i)

$$e^{A^\dagger} = (e^A)^\dagger, \tag{3.6.2}$$

and

$$e^{TAT^{-1}} = Te^AT^{-1}. \tag{3.6.3}$$

(ii) If the matrices A and B commute, then

$$e^{A+B} = e^Ae^B. \tag{3.6.4}$$

(iii)

$$e^Ae^{-A} = I. \tag{3.6.5}$$

(iv) The exponential maps a neighbourhood of 0 in the space of

$n \times n$ matrices, diffeomorphically, onto a neighbourhood of the identity matrix, I.

Thus, if A is anti-Hermitian: $A^\dagger = -A$, it follows from (i) and (iii) that e^A is unitary:

$$e^A e^{-A} = e^A e^{A^\dagger} = e^A (e^A)^\dagger = I. \tag{3.6.6}$$

Conversely, if e^A is unitary, and A belongs to a sufficiently small neighbourhood of 0, the properties (i), (iii) and (iv) imply that $A + A^\dagger = 0$. By virtue of these elementary calculations one can prove two non-trivial results.

(v) The unitary group $U(n)$ is a smooth submanifold of the space of $n \times n$ matrices.

(vi) The tangent space $TU(n)_I$ can be identified with the space of $n \times n$ anti-Hermitian matrices.

This makes it possible to identify also the Lie algebra of $U(n)$, say $u(n)$, with the space of anti-Hermitian matrices, because any tangent vector at I admits a unique extension to a left-invariant vector field on $U(n)$. Moreover, the bracket product of left-invariant vector fields corresponds to the product $AB - BA$ of matrices (e.g. Isham 1989). Note that, for each anti-Hermitian matrix A the correspondence $t \rightarrow e^{tA}$ defines a one-parameter sub-group of $U(n)$, and hence defines a geodesic. But then the map $\exp : TU(n)_I \rightarrow U(n)$ defined by exponentiating matrices coincides with the exponential map in the theory of geodesics, say γ, on Riemannian manifolds:

$$\gamma : [0,1] \rightarrow M, \tag{3.6.7a}$$

$$\gamma(0) = q, \; \frac{d\gamma}{dt}(0) = v, \tag{3.6.7b}$$

$$\gamma(t) = \exp_q(tv). \tag{3.6.8}$$

Starting from the matrices $A, B \in u(n)$, one can define

$$\langle A, B \rangle \equiv \mathrm{Re} \sum_{i,j} A_{ij} \overline{B}_{ij}, \tag{3.6.9}$$

which is a scalar product on $u(n)$. This determines, in turn, a unique left-invariant Riemannian metric on $U(n)$. Note also that each $S \in U(n)$ determines an inner automorphism $X \rightarrow SXS^{-1}$

of the unitary group. The induced linear map from $TU(n)_I$ to $TU(n)_I$ is called the *adjoint action*, and denoted by $Ad(S)$. Interestingly, the scalar product (3.6.9) is invariant under each adjoint action, because, on setting

$$A_1 \equiv Ad(S)A, \tag{3.6.10}$$

$$B_1 \equiv Ad(S)B, \tag{3.6.11}$$

one finds

$$A_1 B_1^\dagger = SAS^{-1}(SBS^{-1})^\dagger = SAB^\dagger S^{-1}, \tag{3.6.12}$$

which implies that

$$\mathrm{tr}(A_1 B_1^\dagger) = \mathrm{tr}\Big(SAB^\dagger S^{-1}\Big) = \mathrm{tr}(AB^\dagger), \tag{3.6.13}$$

and hence

$$\langle A_1, B_1 \rangle = \langle A, B \rangle. \tag{3.6.14}$$

But then the metric on $U(n)$ determined by (3.6.9) is both left- and right-invariant.

Given a matrix $A \in u(n)$, there exists $T \in U(n)$ such that TAT^{-1} is a diagonal matrix,

$$TAT^{-1} = \mathrm{diag}\Big(ia_1, ..., ia_n\Big), \tag{3.6.15}$$

the a_i being real. Moreover, *for any* $S \in U(n)$, one can find a $T \in U(n)$ such that

$$TST^{-1} = \mathrm{diag}\Big(e^{ia_1}, ..., e^{ia_n}\Big), \tag{3.6.16}$$

the a_i being, again, real. This is a direct proof that the map $\exp : u(n) \rightarrow U(n)$ is onto.

A similar analysis may be performed for $SU(n)$, i.e. the group of unitary $n \times n$ matrices of unit determinant. A simple but fundamental identity holds,

$$\det(e^A) = e^{\mathrm{tr}A}. \tag{3.6.17}$$

Thus, the Lie algebra of $SU(n)$, denoted by $su(n)$, consists of all anti-Hermitian traceless matrices:

$$A + A^\dagger = 0, \ \mathrm{tr}A = 0. \tag{3.6.18}$$

Consider now the set of all geodesics in $U(n)$ from I to $-I$. This means that one looks for all $A \in TU(n)_I = u(n)$ such that $e^A = -I$. Such matrices may or may not be in diagonal form. In the latter case, diagonalization may be achieved by means of some $T \in U(n)$ such that TAT^{-1} is diagonal. At this stage, one has

$$e^{TAT^{-1}} = Te^{A}T^{-1} = T(-I)T^{-1} = -I. \qquad (3.6.19)$$

This means that one can safely assume that A is already in diagonal form,

$$A = \text{diag}\Big(ia_1, ..., ia_n\Big), \qquad (3.6.20)$$

which leads to

$$e^{A} = \text{diag}\Big(e^{ia_1}, ..., e^{ia_n}\Big). \qquad (3.6.21)$$

The latter property is consistent with the condition $e^{A} = -I$ if and only if, in Eq. (3.6.20), one can write $a_l = k_l \pi, \forall l = 1, 2, ..., n$, the $\{k_l\}$ being a set of odd integers.

Now we are in a position to understand why the space, say Ω, of all minimal geodesics in $U(n)$ from I to $-I$ may be identified with the space of all vector sub-spaces of C^n. The argument is as follows. First, since the length of the geodesic $t \rightarrow e^{tA}$ from $t = 0$ to $t = 1$ is $\mid A \mid = \sqrt{\text{tr} AA^{\dagger}}$, the length of the geodesic determined by A is $\pi\sqrt{\sum_{i=1}^{n} k_i^2}$, because we have just proved that

$$e^{A} = \text{diag}\Big(e^{ik_1\pi}, ..., e^{ik_n\pi}\Big). \qquad (3.6.22)$$

One can thus say that A determines a minimal geodesic if and only if each k_l equals ± 1, and the corresponding length is then $\pi\sqrt{n}$. But such an A may be viewed as a linear map of C^n into C^n, and hence is completely specified by the vector spaces

$$\text{Eigen}(i\pi) \equiv \{u \in C^n : Au = i\pi u\}, \qquad (3.6.23)$$

$$\text{Eigen}(-i\pi) \equiv \{u \in C^n : Au = -i\pi u\}. \qquad (3.6.24)$$

Moreover, since C^n may be expressed by the direct sum

$$C^n = \text{Eigen}(i\pi) \oplus \text{Eigen}(-i\pi),$$

the matrix A turns out to be completely determined by $\text{Eigen}(i\pi)$, which is an *arbitrary* sub-space of C^n. By virtue of the arbitrariness of $\text{Eigen}(i\pi)$, the identification of Ω with the set of all vector sub-spaces of C^n is obtained.

However, such an identification does not provide a mathematical object that is easily handled, the reason being that Ω has, by definition, components of varying dimensions. To overcome this difficult point, one considers $SU(n)$ instead of $U(n)$, and one sets $n = 2m$. But now the restriction (see (3.6.18))

$$\sum_{i=1}^{2m} a_i = 0 \quad \text{with} \quad a_i = \pm\pi, \tag{3.6.25}$$

forces Eigen$(i\pi)$ to being an m-dimensional (otherwise arbitrary) vector sub-space of C^{2m}. A non-trivial result is then obtained from this analysis (Milnor 1963a).

P.3.6.1 The space of minimal geodesics from I to $-I$ in the special unitary group $SU(2m)$ is *homeomorphic* to the manifold consisting of all m-dimensional vector sub-spaces of C^{2m} (such a manifold is usually denoted by $G_m(C^{2m})$, and is a complex Grassmann manifold).

Moreover, the following properties are known to hold.

P.3.6.2 Every non-minimal geodesic from I to $-I$ in $SU(2m)$ has index $\geq 2m+2$.

P.3.6.3 If the space Ω^d of minimal geodesics from p to q is a topological manifold, and if every non-minimal geodesic from p to q has index $\geq \lambda_0$, the homotopy group $\pi_i(\Omega^d)$ is isomorphic to $\pi_{i+1}(M)$, for $i \in [0, \lambda_0 - 2]$.

By virtue of the properties P.3.6.1–P.3.6.3, a theorem proved by Bott is obtained.

P.3.6.4 The inclusion map $G_m(C^{2m}) \rightarrow \Omega\Big(SU(2m; I, -I)\Big)$ induces isomorphisms of homotopy groups in dimensions $\leq 2m$. This leads to

$$\pi_i G_m(C^{2m}) \cong \pi_{i+1} SU(2m), \tag{3.6.26}$$

for $i \leq 2m$.

On the other hand, homotopy theory provides a further set of useful isomorphisms.

P.3.6.5 The following isomorphisms of homotopy groups hold:

$$\pi_i G_m(C^{2m}) \cong \pi_{i-1} U(m) \quad i \leq 2m, \tag{3.6.27}$$

$$\pi_{i-1}U(m) \cong \pi_{i-1}U(m+1) \cong \pi_{i-1}U(m+2) \cong \ldots \quad i \leq 2m, \tag{3.6.28}$$

$$\pi_j U(m) \cong \pi_j SU(m) \quad \forall j \neq 1. \tag{3.6.29}$$

By virtue of the isomorphisms (3.6.25)–(3.6.29) one then finds the isomorphism which constitutes the Bott periodicity theorem for the unitary group:

$$\pi_{i-1}U(m) \cong \pi_i G_m(C^{2m}) \cong \pi_{i+1}SU(2m) \cong \pi_{i+1}U(m), \tag{3.6.30a}$$

if the integer $i \in [1, 2m]$. In other words, one has

$$\pi_{i-1}U \cong \pi_{i+1}U \quad \forall i \geq 1. \tag{3.6.30b}$$

As an example, one can consider $U(1)$, which is a circle. Hence one finds

$$\pi_0 U = \pi_0 U(1) = 0, \tag{3.6.31}$$

$$\pi_1 U = \pi_1 U(1) \cong Z, \tag{3.6.32}$$

$$\pi_2 U = \pi_2 SU(2) = 0, \tag{3.6.33}$$

$$\pi_3 U = \pi_3 SU(2) \cong Z. \tag{3.6.34}$$

Thus, the stable homotopy groups $\pi_i U$ of the unitary groups are periodic with period 2. In particular, the groups

$$\pi_0 U \cong \pi_2 U \cong \pi_4 U \cong \ldots$$

are zero, whereas the groups

$$\pi_1 U \cong \pi_3 U \cong \pi_5 U \cong \ldots$$

are infinite cyclic.

Now we move towards the formulation of the periodicity theorem for the orthogonal group. To begin, recall that the orthogonal group $O(n)$ may be viewed as the smooth sub-group of the unitary group consisting of all linear maps $T : \Re^n \rightarrow \Re^n$ which preserve the usual inner product on $\Re^n$. A key element of the analysis is the concept of *complex structure.* For this purpose, recall that a complex structure J on $\Re^n$ is a linear map $J : \Re^n \rightarrow \Re^n$, belonging to the orthogonal group, and such that its square is minus the identity: $J^2 = -I$. The space of all complex structures on $\Re^n$ is here denoted by $\Omega_1(n)$, following Milnor (1963a). A first interesting property for us is that the space of minimal geodesics from I to $-I$ on $O(n)$ is homeomorphic to $\Omega_1(n)$. Indeed, since $O(n)$ can be identified with the group of $n \times n$ orthogonal matrices, its tangent space $o(n) \equiv TO(n)_I$ can be identified with the space of $n \times n$ anti-symmetric matrices. Any geodesic γ with initial point I: $\gamma(0) = I$, can be uniquely expressed as

$$\gamma(t) = e^{\pi t A}, \tag{3.6.35}$$

for some $A \in TO(n)_I$. If $n = 2m$, by virtue of the anti-symmetry of A there exists an element $T \in O(n)$ such that

$$TAT^{-1} = \mathrm{diag}(b_1, ..., b_m), \tag{3.6.36}$$

where, with our notation, b_l is the 2×2 matrix

$$b_l \equiv \begin{pmatrix} 0 & a_l \\ -a_l & 0 \end{pmatrix}, \tag{3.6.37}$$

the a_l being non-negative numbers $\forall l = 1, ..., m$. Moreover, the matrix $Te^{\pi A}T^{-1}$ is found to be

$$Te^{\pi A}T^{-1} = \mathrm{diag}(\tilde{b}_1, ..., \tilde{b}_m), \tag{3.6.38}$$

where

$$\tilde{b}_l \equiv \begin{pmatrix} \cos \pi a_l & \sin \pi a_l \\ -\sin \pi a_l & \cos \pi a_l \end{pmatrix}, \tag{3.6.39}$$

$\forall l = 1, ..., m$. This implies that only upon choosing all the a_l as equal to odd integers, the exponentiated matrix $e^{\pi A}$ can coincide with minus the identity.

The inner product

$$\langle A, A \rangle = 2 \sum_{i=1}^{m} a_i^2, \tag{3.6.40}$$

and hence the geodesic $\gamma(t) = e^{\pi t A}$ from I to $-I$ is minimal if and only if $a_l = 1, \forall l = 1, ..., m$. Under this condition, and setting

$$\tilde{J} \equiv \begin{pmatrix} 0 & 1 \\ -1 & 0 \end{pmatrix}, \tag{3.6.41}$$

one finds

$$A^2 = T^{-1}\Big(\mathrm{diag}(\tilde{J}, ..., \tilde{J})\Big)^2 T = -I, \tag{3.6.42}$$

and hence A is, itself, a complex structure. Conversely, let us begin by considering a generic complex structure, say again J. Since, by definition, complex structures are linear maps belonging to the orthogonal group, one has

$$JJ^t = I. \tag{3.6.43}$$

On the other hand, by definition,

$$J^2 = JJ = -I. \tag{3.6.44}$$

By comparison with Eq. (3.6.43), one finds $J^t = -J$, and hence J is anti-symmetric, so that TJT^{-1} takes the form described by (3.6.36) and (3.6.37). Moreover, the property (3.6.44) implies that $a_l = 1, \forall l = 1, ..., m$. This leads to $e^{\pi J} = -I$. The homeomorphism between $\Omega_1(n)$ and the space of minimal geodesics from I to $-I$ on $O(n)$ is therefore proved.

Moreover, one can prove that any non-minimal geodesic from I to $-I$ in $O(2m)$ has index $\geq 2m-2$, and that the space $\Omega_1(n)$ is a manifold. Thus, using the result P.3.6.3, one obtains yet another theorem due to Bott.

P.3.6.6 The inclusion map $\Omega_1(n) \rightarrow \Omega O(n)$ induces isomorphisms of homotopy groups in dimensions $\leq n-4$, so that

$$\pi_i \Omega_1(n) \cong \pi_{i+1} O(n), \tag{3.6.45}$$

for $i \leq n-4$.

The next step consists in the introduction of some fixed complex structures on $\Re^n$, say $J_1, ..., J_{k-1}$, which anti-commute, in that

$$J_r J_s + J_s J_r = 0 \quad \text{if } r \neq s. \tag{3.6.46}$$

One also assumes that there exists at least another complex structure, say J, which anti-commutes with $J_l, \forall l = 1, ..., k-1$. One then defines $\Omega_k(n)$ as the set of all complex structures J on $\Re^n$ which anti-commute with the fixed structures $J_l, \forall l = 1, ..., k-1$. By construction, one has

$$\Omega_k(n) \subset \Omega_{k-1}(n) \subset ... \subset \Omega_1(n) \subset O(n), \tag{3.6.47}$$

and each $\Omega_k(n)$ is a compact set. At this stage, one is led to define

$$\Omega_0(n) \equiv O(n). \tag{3.6.48}$$

Interestingly, each $\Omega_k(n)$ is a smooth sub-manifold of $O(n)$, and each geodesic in $\Omega_k(n)$ is also a geodesic in $O(n)$.

Note now that Je^A is a complex structure if and only if A anti-commutes with J. In fact, if A anti-commutes with J, then

$$J^{-1}AJ = -J^{-1}JA = -A,$$

and hence

$$I = e^{-A}e^A = e^{J^{-1}AJ}\, e^A = J^{-1}e^A J e^A.$$

This leads to $Je^A = e^{-A}J$, which implies $(Je^A)^2 = -I$. Conversely, if $(Je^A)^2 = -I$, the above analysis shows that

$$e^{J^{-1}AJ}\, e^A = I.$$

Since A is taken to be 'small', this equality implies that $J^{-1}AJ = -A$, which, upon multiplication of both sides by J on the left, shows that A anti-commutes with J.

Moreover, Je^A anti-commutes with the complex structures J_l, for all $l = 1, ..., k-1$, if and only if A commutes with $J_l, \forall l = 1, ..., k-1$. By virtue of these elementary properties, a neighbourhood of J in $\Omega_k(n)$ is given by all points Je^A, where A ranges over all 'small' matrices in a linear sub-space of the Lie algebra $o(n)$. This implies that each geodesic in $\Omega_k(n)$ is also a geodesic in $O(n)$.

What is very important is the possibility to imbed $\Omega_k(n)$ in $\Omega_k(n+n')$. For this purpose, one may choose some fixed anti-commuting complex structures $J'_1, ..., J'_k$ on $\Re^{n'}$, so that each $J \in \Omega_k(n)$ determines a complex structure $J \oplus J'_k$ on $\Re^n \oplus \Re^{n'}$, which anti-commutes with $J_\alpha \oplus J'_\alpha$ for $\alpha = 1, ..., k-1$. Denoting by Ω_k the direct limit with the fine topology, as $n \rightarrow \infty$, of the spaces $\Omega_k(n)$,

$$\Omega_k \equiv \lim_{n \rightarrow \infty} \Omega_k(n), \tag{3.6.49}$$

the space

$$\mathcal{O} \equiv \Omega_0 \tag{3.6.50}$$

is called the *infinite orthogonal group*. As far as the spaces $\Omega_k(n)$ are concerned, after setting $n \equiv 16r$, one should bear in mind that $\Omega_8(16r)$ is diffeomorphic to the orthogonal group $O(r)$. For $k > 8$, it can be shown that $\Omega_k(16r)$ is diffeomorphic to $\Omega_{k-8}(r)$, but we will not need to go beyond the case $k = 8$. Remarkably, on taking the limit $r \rightarrow \infty$ in the diffeomorphism between $\Omega_8(16r)$ and $O(r)$, one finds that, with the notation (3.6.49), Ω_8 is homeomorphic to the infinite orthogonal group $\mathcal{O}$. Moreover, homotopy theory ensures that the limit map is a homotopy equivalence. The joint effect of all these properties is the Bott periodicity theorem for the infinite orthogonal group $\mathcal{O}$, according to which $\mathcal{O}$ has the same homotopy type as its own 8th loop space. Hence one finds the isomorphism

$$\pi_i \mathcal{O} \cong \pi_{i+8} \mathcal{O}, \tag{3.6.51}$$

for $i \geq 0$.

All these results may be re-expressed by using the symbol of elliptic operators. To understand the key features, let us consider an elliptic operator on $\Re^n$ with constant coefficients, say P, which acts on C^m-valued functions. The symbol of P, denoted by $\sigma_P(\xi)$, provides a map

$$\sigma_P : S^{n-1} \to GL(m, C).$$

For example, the restricted Dirac operator on $\Re^{2k}$, say B, yields the map

$$\sigma_B : S^{2k-1} \to GL(2^{k-1}, C).$$

The Bott periodicity theorem states that this is a stable generator, in that

(i) the map

$$\pi_i(GL(m, C)) \to \pi_i(GL(m+1, C))$$

is an isomorphism for $m > \frac{i}{2}$;

(ii) the limit group $\pi_i(GL(\infty))$ is 0 if i is even, and coincides with Z, generated by σ_B, if i is odd.

A global version of the Bott theorem, resulting from the local form, is also available. According to the global Bott theorem, *the symbol of the Dirac operator on a compact manifold is a generator*, in a suitable sense, *of all elliptic symbols.* More precisely, let M be a compact, oriented, $2k$-dimensional manifold endowed with a spin-structure (see section 1.3). We know from section 2.3 that an elliptic differential operator

$$P : C^\infty(V, M) \to C^\infty(W, M)$$

has a leading symbol, say σ_P, which is an isomorphism

$$\pi^* V \to \pi^* W,$$

where $\pi : S(M) \to M$ is the projection map of the unit sphere-bundle. A *trivial symbol* is a symbol for which $V = W$ and σ is the identity map. Two symbols are said to be *equivalent* if they become homotopic after adding trivial symbols.

The global Bott theorem can be then stated by saying that *every symbol is equivalent to the symbol of B_V, for some V*, where B_V is the restricted Dirac operator, extended so as to act on the tensor product $E^+ \otimes V$. For index problems, the interest of this property lies in the fact that *two elliptic operators whose symbols are equivalent turn out to have the same index.* Thus, if one is able to find the index of B_V for all V, one can indeed obtain an index formula for all elliptic operators on M.

3.7 K-theory

The global Bott theorem and the global theory of the Dirac operator find their natural formulation within the framework of K-theory. The formalism is as abstract as powerful, and some key steps are as follows (Atiyah 1967).

We are interested in complex vector bundles over spaces, say M, which can be topological spaces, or compact Hausdorff, or closed Riemannian manifolds, or spaces of any nature. The set $Vect(M)$ of isomorphism classes of vector bundles over M has the structure of an Abelian semi-group (this property holds independently of the precise nature of M), where the additive structure is defined by direct sum. If A is *any* Abelian semi-group (recall that a semi-group is a set of elements $\{a, b, c, ...\}$ with a binary operation of product, say ab, such that $(ab)c = a(bc)$), one can associate to A an Abelian group, $K(A)$, with the following property: there is a semi-group homomorphism

$$\alpha : A \rightarrow K(A) \tag{3.7.1}$$

such that, if G is *any* group, and if

$$\gamma : A \rightarrow G \tag{3.7.2}$$

is *any* semi-group homomorphism, there is a *unique* homomorphism

$$\kappa : K(A) \rightarrow G \tag{3.7.3}$$

such that

$$\gamma = \kappa \, \alpha. \tag{3.7.4}$$

If such a $K(A)$ exists, it must be unique. Note that the order of factors on the right-hand side of Eq. (3.7.4) is of the right sort: first, α maps A into $K(A)$, then κ maps $K(A)$ into G. Altogether, this is a map of A into G. What is non-trivial is the *uniqueness* property of the homomorphism.

To define the group $K(A)$, one starts from $F(A)$, the free Abelian group generated by the elements of A. Second, one considers $E(A)$, the sub-group of $F(A)$ generated by those elements of the form

$$a + a' - (a \oplus a'),$$

where $\oplus$ is the addition in A, and $a, a' \in A$. Then

$$K(A) \equiv F(A)/E(A) \tag{3.7.5}$$

is the desired Abelian group. If A possesses a multiplication which is distributive over the addition of A, then $K(A)$ is a ring.

Now if M is a space, one writes

$$K(M) \equiv K(\mathit{Vect}(M)) \tag{3.7.6}$$

for the ring $K(\mathit{Vect}(M))$. Moreover, if $E \in \mathit{Vect}(M)$, one writes $[E]$ for the image of E in $K(M)$. One can prove that every element of $K(M)$ is of the form $[E] - [F]$, where E, F are bundles over M. Let now T be a bundle such that $F \oplus T$ is trivial. On writing $\underline{n}$ for the trivial bundle of dimension n, this means that

$$F \oplus T = \underline{n}. \tag{3.7.7}$$

One can thus obtain a very useful formula for a generic element of $K(M)$:

$$[E] - [F] = [E] + [T] - ([F] + [T]) = [E \oplus T] - [\underline{n}]. \tag{3.7.8}$$

In other words, every element of $K(M)$ is of the form $[H] - [\underline{n}]$. If two bundles become equivalent when a suitable trivial bundle is added to each of them, the bundles are said to be *stably equivalent*. Thus, $[E] = [F]$ if and only if E and F are stably equivalent.

For a space with base point x_0, one defines

$$\widetilde{K}(M) \equiv \mathrm{Ker}\,\{K(M) \rightarrow K(x_0)\}. \tag{3.7.9}$$

Since $K(\text{point}) \cong Z$, one has the isomorphism

$$K(M) \cong \widetilde{K}(M) \oplus Z. \tag{3.7.10}$$

For a closed sub-space of M, say Y, one defines

$$K(M,Y) \equiv \widetilde{K}(M/Y), \tag{3.7.11}$$

where M/Y is obtained by collapsing Y to a point, taken as the base point. For a locally compact space, one sets

$$K(M) \equiv \widetilde{K}(M^+), \tag{3.7.12}$$

where M^+ is the one-point compactification of M. Another elementary lemma can then be proved.

If $Y \subset M$ is a compact pair, then

$$K(M,Y) \rightarrow K(M) \rightarrow K(Y)$$

is an exact sequence (see (3.B.1)).

The tensor product of vector bundles induces a ring structure on $K(M)$, and $K(Y,M)$ becomes a module on $K(M)$. The identity map

$$S^1 \to U(1) \subset GL(1,\mathcal{C})$$

defines a line-bundle, say H, on $S^2 = \mathcal{C}^+$. Let $b \in K(\mathcal{C})$ be the class of $H-1$. The local form of the Bott periodicity theorem is therefore (Atiyah 1975a) as follows.

Theorem 3.7.1 Multiplication by b provides an isomorphism

$$\beta : K(M) \to K(\mathcal{C} \times M).$$

On taking $M = \mathcal{C}^n$ and applying induction for $n \geq 0$, one finds that $K(\mathcal{C}^n) \cong Z$. Moreover, the symbol of the restricted Dirac operator, B, is eventually identified with the element b^n.

In the analysis of global properties, one first proves that an elliptic symbol σ defines an element $[\sigma]$ of $K(T^*(M))$. The global Bott theorem is then the following.

Theorem 3.7.2 For an even-dimensional spin-manifold, multiplication by σ_B provides the isomorphism

$$K(M) \cong K(T^*(M)).$$

In particular, if the manifold M is parallelizable, theorem 3.7.2 reduces to theorem 3.7.1. In the general case, one takes a covering of M, on which $T^*(M)$ is trivial, and one applies induction by using the long exact sequence which extends

$$K(M,Y) \to K(M) \to K(Y).$$

This long exact sequence involves higher-order groups, denoted by $K^{-n}(M)$, defined, for $n \geq 0$, by means of

$$K^{-n}(M) \equiv K(\Re^n \times M). \tag{3.7.13}$$

One may wonder what can be said about manifolds which are not spin-manifolds. Indeed, if M is not a spin-manifold one cannot define a Dirac operator. One can, however, build a signature operator (see section 2.3), say A. Locally, the symbol of A is equivalent to a multiple of the symbol of B (Atiyah 1975a):

$$\sigma_A = 2^k \, \sigma_B. \tag{3.7.14}$$

Appendix 3.A

An elliptic boundary-value problem consists of the pair (P, B), where, for a given vector bundle V over a Riemannian manifold M, P is a differential operator of order d on V, and B is a boundary operator (see (5.1.18)). The *strong ellipticity* of (P, B) is defined in terms of the eigenvalue equation for the leading symbol of P, jointly with an asymptotic condition on the solution of this equation. For us to be able to define these concepts, it is necessary to start from the structure which makes it possible to split the tangent bundle of M in the form of a direct sum (Gilkey 1995)

$$T(M) = T(\partial M) \oplus T([0, \delta)). \tag{3.A.1}$$

Indeed, following Gilkey (1995), one can use the inward geodesic flow to identify a neighbourhood of ∂M with the *collar*

$$\mathcal{C} \equiv \partial M \times [0, i(M)), \tag{3.A.2}$$

$i(M)$ being the injectivity radius. If $(y_1, ..., y_{m-1})$ are local coordinates on the boundary, and if x_m is the geodesic distance to the boundary, a system of local coordinates on the collar is then $(y_1, ..., y_{m-1}, x_m)$. The normal derivatives $D_m^k \left(f \mid_{\mathcal{C}} \right)$ are well defined after identifying $V \mid_{\mathcal{C}}$ with $V \times [0, i(M))$. We now need to define d-graded vector bundles (d being the order of P) and their auxiliary bundles. For this purpose, following Gilkey (1995), we say that a *d-graded vector bundle* is a vector bundle, say U, endowed with a decomposition into d sub-bundles

$$U = U_0 \oplus ... \oplus U_{d-1}. \tag{3.A.3}$$

Moreover, we consider W, defined by

$$W \equiv [V]_{\partial M} \oplus ... \oplus [V]_{\partial M}, \tag{3.A.4}$$

i.e. the d-graded vector bundle for boundary data, and define (Gilkey 1995)

$$f_i \equiv \left[D^i{}_m (f \mid_{\mathcal{C}}) \right]_{\partial M}. \tag{3.A.5}$$

The boundary data map is then a map

$$\overline{\gamma} : C^\infty(V) \rightarrow C^\infty(W)$$

such that

$$\overline{\gamma}(f) \equiv (f_0, ..., f_{d-1}). \tag{3.A.6}$$

Further to this, one has to take into account an *auxiliary* d-graded vector bundle over the boundary, say Y, whose dimension satisfies

$$\dim(Y) = \frac{d}{2}\dim(V). \tag{3.A.7}$$

If $B : C^\infty(W) \rightarrow C^\infty(Y)$ is a tangential differential operator defined on ∂M, one can decompose it as follows:

$$B \equiv B_{ij}, \tag{3.A.8a}$$

$$B_{ij} : C^\infty(W_i) \rightarrow C^\infty(Y_j), \tag{3.A.8b}$$

with $\mathrm{ord}(B_{ij}) \leq j - i$. Sections of $C^\infty(W_i)$ arise from taking the normal derivative of order i.

All these geometric objects make it possible to define the *d-graded leading symbol* of B as follows (Gilkey 1995):

$$\sigma_g(B)_{ij}(y, \zeta) \equiv \sigma_L(B_{ij})(y, \zeta) \text{ if } \mathrm{ord}(B_{ij}) = j - i, \tag{3.A.9a}$$

$$\sigma_g(B)_{ij}(y, \zeta) \equiv 0 \text{ if } \mathrm{ord}(B_{ij}) < j - i. \tag{3.A.9b}$$

With the notation described so far, the boundary condition is expressed by the equation

$$B \, \overline{\gamma} \, f = 0. \tag{3.A.10}$$

To define *strong ellipticity* of the pair (P, B) one assumes that P is a differential operator on V of order d, with elliptic leading symbol, say p_d. If $\mathcal{K}$ is a cone containing 0 and contained in $\mathbf{C}$, and such that, for $\xi \neq 0$,

$$\mathrm{Spec}(p_d(x, \xi)) \subset \mathcal{K}^c, \tag{3.A.11}$$

one studies, on the boundary, the equation

$$p_d(y, 0, \zeta, D_r) f(r) = \lambda \, f(r), \tag{3.A.12}$$

subject to the asymptotic condition

$$\lim_{r \rightarrow \infty} f(r) = 0, \tag{3.A.13}$$

and under the assumption that $(0, 0) \neq (\zeta, \lambda) \in T^*(\partial M) \times \mathcal{K}$. By definition, the pair (P, B) is said to be *strongly elliptic* with respect to the cone $\mathcal{K}$ if, *for any* w belonging to the auxiliary d-graded vector bundle Y, there exists a *unique* solution of the problem described by Eqs. (3.A.12) and (3.A.13), with (Gilkey 1995)

$$\sigma_g(B)(y, \zeta) \overline{\gamma}(f) = w. \tag{3.A.14}$$

The cone $\mathcal{K}$ is normally assumed to coincide with the set of complex numbers minus $\Re^+$ (or minus $\Re$ and $\Re^+$).

For example, if P is an operator of Laplace type on $C^\infty(V)$, one finds (Gilkey 1995)

$$p_d(y,0,\zeta,D_r)f(r) = -\partial_r^2 + |\zeta|^2, \tag{3.A.15}$$

and hence the solutions of Eq. (3.A.12) read

$$f(r) = w_0 e^{\mu r} + w_1 e^{-\mu r}, \tag{3.A.16}$$

where $\mu \equiv \sqrt{|\zeta|^2 - \lambda}$. Bearing in mind that $\mathrm{Re}(\mu) > 0$, the asymptotic condition (3.A.13) picks out the solution of the form

$$\tilde{f}(r) = w_1 e^{-\mu r}. \tag{3.A.17}$$

These properties are useful in the course of proving that P is strongly elliptic with respect to the cone $\mathbf{C} - \Re^+$, when the boundary conditions are a mixture of Dirichlet and Robin conditions (see our Eq. (5.4.2) and Gilkey (1995)). A thorough treatment of boundary-value problems, with emphasis on the analytic approach, may be found in Grubb (1996).

Appendix 3.B

This appendix describes briefly some concepts in homotopy theory and in the theory of characteristic classes. It is intended to help the readers who have already attended introductory courses, but do not have enough time to read comprehensive monographs or long review papers.

(i) A very important concept in homotopy theory is that of an exact sequence. A sequence of groups and homomorphisms

$$\ldots \xrightarrow{f_{i-1}} H_i \xrightarrow{f_i} H_{i+1} \xrightarrow{f_{i+1}} H_{i+2} \ldots$$

is called an *exact sequence* if, for all i,

$$\mathrm{Im}\ f_{i-1} = \mathrm{Ker}\ f_i. \tag{3.B.1}$$

A *short* exact sequence is a five-term sequence with trivial end groups, i.e.

$$0 \longrightarrow H \xrightarrow{f} H' \xrightarrow{f'} H'' \longrightarrow 0. \tag{3.B.2}$$

(ii) If m is a complex $k \times k$ matrix, and $Q(m)$ is a polynomial in the components of m, then $Q(m)$ is called a *characteristic polynomial* if

$$Q(m) = Q(h^{-1}\, m\, h) \quad \forall h \in GL(k, C). \tag{3.B.3}$$

The polynomial $Q(m)$ is a symmetric function of the eigenvalues $\{\lambda_1, ..., \lambda_k\}$ of m. The jth symmetric polynomial, say $T_j(\lambda)$, reads

$$T_j(\lambda) = \sum_{i_1 < i_2 < ... i_j} \lambda_{i_1}\, \lambda_{i_2} ... \lambda_{i_j}, \tag{3.B.4}$$

and $Q(m)$ is a polynomial in the $T_j(\lambda)$:

$$Q(m) = \alpha + \beta T_1(\lambda) + \gamma T_2(\lambda) + \delta[T_1(\lambda)]^2 + ... \tag{3.B.5}$$

If, in $Q(m)$, one replaces m by the curvature two-form Ω, one finds that $Q(\Omega)$ is closed.

The *total Chern form* $c(\Omega)$ of a complex vector bundle E over M, with $GL(k, C)$ transition functions and connection ω, is defined in terms of the characteristic polynomial $\det(I + m)$ for Ω:

$$c(\Omega) = \det\left(I + \frac{i}{2\pi}\Omega\right) = \sum_{k=0}^{\infty} c_k(\Omega), \tag{3.B.6}$$

where the various Chern forms in (3.B.6) read

$$c_0(\Omega) = 1, \tag{3.B.7}$$

$$c_1(\Omega) = \frac{i}{2\pi}\text{Tr}\Omega, \tag{3.B.8}$$

$$c_2(\Omega) = \frac{1}{8\pi^2}\Big[\text{Tr}\Omega \wedge \Omega - \text{Tr}\Omega \wedge \text{Tr}\Omega\Big], \tag{3.B.9}$$

plus infinitely many other formulae which can be derived from the expansion of the determinant in (3.B.6). Since $Q(\Omega)$ is closed, one finds that any homogeneous polynomial in the expansion of a characteristic polynomial is closed:

$$d\, c_j(\Omega) = 0. \tag{3.B.10}$$

Thus, the Chern forms, $c_j(\Omega)$, define cohomology classes belonging to $H^{2j}(M)$.

(iii) The *Chern numbers* of a fibre bundle are the numbers found by integrating characteristic polynomials over the manifold. For example, one has

$$C_2(E) \equiv \int_M c_2(\Omega). \tag{3.B.11}$$

(iv) *Pontrjagin classes*, say $p_k(E)$, and *Pontrjagin numbers*, are the counterpart, for *real* vector bundles, of Chern classes and Chern numbers for complex vector bundles.

(v) In the course of studying index theorems, one needs certain combinations of characteristic classes. The first example is provided by the *Chern character*, defined by the invariant polynomial

$$ch(E) \equiv \mathrm{Tr}\exp(im/2\pi) = \sum_{l=0}^{\infty} \frac{1}{l!}\mathrm{Tr}(im/2\pi)^l. \qquad (3.B.12)$$

A deep and simple relation exists between the Chern character and Chern classes, i.e.

$$ch(E) = \sum_{l=1}^{k} \exp(i\Omega_l/2\pi) = k + c_1(E) + \frac{1}{2}\Big(c_1^2 - 2c_2\Big)(E) + ... \qquad (3.B.13)$$

(vi) Yet another deep concept is the one of *genus*, i.e. a combination of characteristic classes that satisfies the Whitney sum property:

$$f(E \oplus E') = f(E)f(E'). \qquad (3.B.14)$$

It is now convenient to set

$$k_l \equiv \frac{i}{2\pi}\Omega_l. \qquad (3.B.15)$$

It is then possible to express the *Todd class* as follows:

$$td(E) \equiv \prod_{l=1}^{k} \frac{x_l}{1-\exp(-x_l)} = 1 + \frac{1}{2}c_1(E) + \frac{1}{12}\Big(c_1^2 + c_2\Big)(E) + ... \qquad (3.B.16)$$

Other relevant examples of genera are given by the *Hirzebruch L-polynomial*, i.e.

$$L(E) \equiv \prod_l \frac{x_l}{\tanh(x_l)}, \qquad (3.B.17)$$

and the $\hat{A}$ polynomial

$$\hat{A}(E) \equiv \prod_l \frac{x_l/2}{\sinh(x_l/2)} = 1 - \frac{1}{24}p_1 + \frac{1}{5760}\Big(7p_1^2 - 4p_2\Big) + ... \qquad (3.B.18)$$

(vii) A further important set of characteristic classes consists of the Stiefel–Whitney classes. Unlike all characteristic classes described previously, they cannot be represented by differential forms in terms of curvature, and are not integral cohomology classes (Milnor and Stasheff 1974, Rennie 1990). By definition, Stiefel–Whitney classes are the Z_2 cohomology classes of a *real* bundle, say E, over M, with k-dimensional fibre:

$$w_i \in H^i(M; Z_2), \; i = 1, ..., n-1. \tag{3.B.19}$$

The *total Stiefel–Whitney class* is defined by

$$w(E) \equiv 1 + \sum_{l=1}^{n} w_l. \tag{3.B.20}$$

The vanishing of the first Stiefel–Whitney class $w_1(T(M))$ provides a necessary and sufficient condition for the orientability of M. The vanishing of the second Stiefel–Whitney class $w_2(T(M))$ is instead a necessary and sufficient condition for the existence of spin-structures (see sections 1.2, 1.3 and Milnor (1963b)).

4
Spectral asymmetry

The investigations in spectral asymmetry and Riemannian geometry by Atiyah, Patodi and Singer begin by studying, for a Riemannian manifold, say M, the relation between the signature of the quadratic form on $H^2(M;\Re)$ and the integral over M of the first Pontrjagin class. It turns out that, if M has a boundary, the desired relation involves also the value at the origin of the η-function obtained from the eigenvalues of a first-order differential operator. The result is indeed an example of index theorem for non-local boundary conditions. After this, a detailed $\eta(0)$ calculation is performed with non-local boundary conditions, and, as a further example, the $\eta(0)$ value is obtained for a first-order operator with periodic boundary conditions. The second part of the chapter is devoted to some physical applications of this scheme. Thus, the one-loop approximation is studied for a massless spin-1/2 field on a flat four-dimensional Euclidean background bounded by two concentric three-spheres, when non-local boundary conditions of the spectral type are imposed. The use of ζ-function regularization shows that the conformal anomaly vanishes, as in the case when the same field is subject to local boundary conditions involving projectors. A similar analysis of non-local boundary conditions can be performed for massless supergravity models on manifolds with boundary, to study their one-loop properties. Moreover, it is shown that the proof of essential self-adjointness for the boundary-value problem can be obtained by means of Weyl's limit point – limit circle criterion, and bearing in mind the properties of continuous potentials which are positive near zero and are bounded on the interval $[1, \infty[$.

4.1 Spectral asymmetry and Riemannian geometry

If M is an oriented, m-(even) dimensional Riemannian manifold, the Gauss–Bonnet–Chern theorem provides a first relevant example of a deep formula relating cohomological invariants with curvature. In intrinsic language, the Euler number χ of M, defined as an alternating sum of Betti numbers:

$$\chi \equiv \sum_{p=0}^{n} (-1)^p B_p, \tag{4.1.1}$$

reads (Chern 1944, 1945, Greub *et al.* 1973, Dowker and Schofield 1990)

$$\chi = \int_M \Lambda + \int_{\partial M} X^* \Pi, \tag{4.1.2a}$$

where the vector field X can be chosen as any extension of the normal vector field $\mathbf{n}$ on the boundary ∂M. Moreover, Π is an $(m-1)$-form on the unit tangent bundle of M, the sphere-bundle $S(M)$, and Λ is an m-form such that $\pi^* \Lambda$ is the Pfaffian of the curvature matrix, where π^* maps the cohomology of M into that of $S(M)$. For example, in the two-dimensional case, Eq. (4.1.2a) reduces to

$$\chi = \frac{1}{4\pi} \int_M R + \frac{1}{2\pi} \int_{\partial M} \mathrm{Tr} K, \tag{4.1.2b}$$

where R is the trace of the Ricci tensor, and K is the second fundamental form of the boundary of M (see section 5.2).

When M is four-dimensional, in addition to (4.1.2a) there is another formula which relates cohomological invariants with curvature. In fact, it is known that the signature (i.e. number of positive eigenvalues minus number of negative eigenvalues) of the quadratic form on $H^2(M; \Re)$ given by the cup-product (see section 2.3 and Atiyah (1975a)) is expressed, for manifolds without boundary, by

$$\mathrm{sign}(M) = \frac{1}{3} \int_M p_1. \tag{4.1.3}$$

In (4.1.3), p_1 is the differential four-form which represents the first Pontrjagin class, and is equal to $(2\pi)^{-2}$ $\mathrm{Tr}(R^2)$, where R is the curvature matrix. However, (4.1.3) does not hold in general for manifolds with boundary, so that one has

$$\text{sign}(M) - \frac{1}{3}\int_M p_1 = f(Y) \neq 0, \tag{4.1.4}$$

where $Y \equiv \partial M$. Thus, if M' is another manifold with the same boundary, i.e. such that $Y = \partial M'$, one has

$$\text{sign}(M) - \frac{1}{3}\int_M p_1 = \text{sign}(M') - \frac{1}{3}\int_{M'} p_1'. \tag{4.1.5}$$

Hence one is looking for a continuous function f of the metric on Y such that $f(-Y) = -f(Y)$. Atiyah *et al.* (1975) were able to prove that $f(Y)$ is a spectral invariant, evaluated as follows. One looks at the Laplace operator $\triangle$ acting on forms as well as on scalar functions. This operator $\triangle$ is the square of the self-adjoint first-order operator $B \equiv \pm\left(d * - * d\right)$, where d is the exterior-derivative operator, and $*$ is the Hodge-star operator mapping p-forms to $(l-p)$-forms in l dimensions. Thus, if λ is an eigenvalue of B, the eigenvalues of $\triangle$ are of the form λ^2. However, the eigenvalues of B can be both negative and positive. One takes this property into account by defining the η-function

$$\eta(s) \equiv \sum_{\lambda \neq 0} d(\lambda)(\text{sign}(\lambda))|\,\lambda\,|^{-s}, \tag{4.1.6}$$

where $d(\lambda)$ is the multiplicity of the eigenvalue λ. Note that, since B involves the $*$ operator, in reversing the orientation of the boundary Y we change B into $-B$, and hence $\eta(s)$ into $-\eta(s)$. The main result of Atiyah *et al.*, in its simplest form, states therefore that (Atiyah *et al.* 1975, Atiyah 1975b)

$$f(Y) = \frac{1}{2}\eta(0). \tag{4.1.7}$$

Now, for a manifold M with boundary Y, if one tries to set up an elliptic boundary-value problem for the signature operator of section 2.3, one finds that there is no local boundary condition for this operator for which the ellipticity of the boundary-value problem is obtained. For global boundary conditions, however, expressed by the vanishing of a given integral evaluated on Y, one has a good elliptic theory and a finite index. Thus, one has to consider the theorem expressed by (4.1.7) within the framework of index theorems for global boundary conditions. Atiyah *et al.* (1975) were also able to derive the relation between the index of the Dirac operator on M with a global boundary condition and $\eta(0)$, where η is the η-function of the Dirac operator on the boundary of M (cf. (4.4.4) and (4.4.5)).

4.2 $\eta(0)$ calculation

Following Atiyah *et al.* (1975), we now evaluate $\eta(0)$ in a specific example. Let Y be a closed manifold, E a vector bundle over Y and $A : C^\infty(E, Y) \rightarrow C^\infty(E, Y)$ a self-adjoint, elliptic, first-order differential operator. By virtue of this hypothesis, A has a discrete spectrum with real eigenvalues λ and eigenfunctions ϕ_λ. Let P denote the projection of $C^\infty(E, Y)$ onto the space spanned by the ϕ_λ for $\lambda \geq 0$. We now form the product $Y \times \Re^+$ of Y with the half-line $u \geq 0$ and consider the operator

$$D \equiv \frac{\partial}{\partial u} + A, \tag{4.2.1}$$

acting on sections $f(y, u)$ of E lifted to $Y \times \Re^+$ (still denoted by E). Clearly D is elliptic and its *formal* adjoint is

$$D^* \equiv -\frac{\partial}{\partial u} + A. \tag{4.2.2}$$

The following boundary condition is imposed for D:

$$Pf(\cdot, 0) = 0. \tag{4.2.3}$$

This is a *global* condition for the boundary value $f(\cdot, 0)$ in that it is equivalent to

$$\int_Y \Big(f(y, 0), \phi_\lambda(y) \Big) = 0 \quad \text{for all } \lambda \geq 0. \tag{4.2.4}$$

Of course, the adjoint boundary condition to (4.2.3) is

$$(1 - P) f(\cdot, 0) = 0. \tag{4.2.5}$$

The naturally occurring second-order self-adjoint operators obtained from D are

$$\Delta_1 \equiv \mathcal{D}^* \mathcal{D}, \tag{4.2.6}$$

$$\Delta_2 \equiv \mathcal{D} \mathcal{D}^*, \tag{4.2.7}$$

where $\mathcal{D}$ is the closure of the operator D on L^2 with domain given by (4.2.3). For $t > 0$, one can then consider the bounded operators $e^{-t\Delta_1}$ and $e^{-t\Delta_2}$. The explicit kernels of these operators (see section 5.2 and appendix 5.A) will be given in terms of the eigenfunctions ϕ_λ of A. For this purpose, consider first Δ_1, i.e. the operator given by

$$-\frac{\partial^2}{\partial u^2}+A^2,$$

with the boundary condition

$$Pf(\cdot,0)=0, \tag{4.2.8a}$$

and

$$(1-P)\left(\frac{\partial f}{\partial u}+Af\right)_{u=0}=0. \tag{4.2.8b}$$

Expansion in terms of the ϕ_λ, so that $f(y,u)=\sum_\lambda f_\lambda(u)\phi_\lambda(y)$, shows that, for each λ, one has to study the operator

$$-\frac{d^2}{du^2}+\lambda^2$$

on $u \geq 0$, with the boundary conditions

$$f_\lambda(0)=0 \quad \text{if} \quad \lambda \geq 0, \tag{4.2.9}$$

$$\left(\frac{df_\lambda}{du}+\lambda f_\lambda\right)_{u=0}=0 \quad \text{if} \quad \lambda<0. \tag{4.2.10}$$

It should be stressed, once more, that (4.2.9) and (4.2.10) are non-local, since they rely on the separation of the spectrum of A into its positive and negative part (cf. section 4.4).

The fundamental solution for

$$\frac{\partial}{\partial t}-\frac{\partial^2}{\partial u^2}+\lambda^2$$

with the boundary condition (4.2.9) is found to be

$$W_A=\frac{e^{-\lambda^2 t}}{\sqrt{4\pi t}}\left[\exp\left(\frac{-(u-v)^2}{4t}\right)-\exp\left(\frac{-(u+v)^2}{4t}\right)\right]. \tag{4.2.11}$$

By contrast, when the boundary condition (4.2.10) is imposed, the use of the Laplace transform leads to (Carslaw and Jaeger 1959)

$$W_B=\frac{e^{-\lambda^2 t}}{\sqrt{4\pi t}}\left[\exp\left(\frac{-(u-v)^2}{4t}\right)+\exp\left(\frac{-(u+v)^2}{4t}\right)\right] + \lambda e^{-\lambda(u+v)}\,\mathrm{erfc}\left[\frac{(u+v)}{2\sqrt{t}}-\lambda\sqrt{t}\right], \tag{4.2.12}$$

where erfc is the (complementary) error function defined by

$$\mathrm{erfc}(x) \equiv \frac{2}{\sqrt{\pi}} \int_x^\infty e^{-\xi^2}\, d\xi. \tag{4.2.13}$$

Thus, the kernel K_1 of $e^{-t\Delta_1}$ at a point $(y, u; z, v; t)$ is obtained as

$$K_1(y, u; z, v; t) = \sum_\lambda W_A \phi_\lambda(y) \overline{\phi_\lambda(z)} \quad \text{if } \lambda \geq 0, \tag{4.2.14}$$

$$K_1(y, u; z, v; t) = \sum_\lambda W_B \phi_\lambda(y) \overline{\phi_\lambda(z)} \quad \text{if } \lambda < 0. \tag{4.2.15}$$

For the operator Δ_2, the boundary conditions for each λ are

$$f_\lambda(0) = 0 \quad \text{if} \quad \lambda < 0, \tag{4.2.16}$$

$$\left(-\frac{df_\lambda}{du} + \lambda f_\lambda \right)_{u=0} = 0 \quad \text{if} \quad \lambda \geq 0. \tag{4.2.17}$$

The fundamental solution for

$$\frac{\partial}{\partial t} - \frac{\partial^2}{\partial u^2} + \lambda^2$$

subject to (4.2.16) is $\widetilde{W}_A = W_A$, while for the boundary conditions (4.2.17) one finds $\widetilde{W}_B = W_B(-\lambda)$. Moreover, by virtue of the inequality

$$\int_x^\infty e^{-\xi^2}\, d\xi < e^{-x^2}, \tag{4.2.18}$$

W_A and W_B are both bounded by

$$F_\lambda(u, v; t) \equiv \left[\frac{e^{-\lambda^2 t}}{\sqrt{\pi t}} + \frac{2 \mid \lambda \mid}{\sqrt{\pi}} e^{-\lambda^2 t} \right] \exp\left(\frac{-(u-v)^2}{4t} \right). \tag{4.2.19}$$

If one now multiplies and divides by $\sqrt{t}$ the second term in square brackets in (4.2.19), application of the inequality

$$x \leq e^{x^2/2}, \tag{4.2.20}$$

to the resulting term $\mid \lambda \mid \sqrt{t}$ shows that the kernel $K_1(y, u; z, v; t)$ of $e^{-t\Delta_1}$ is bounded by

$$G(y, u; z, v; t) \equiv \frac{3}{2\sqrt{\pi t}} \left[\sum_\lambda e^{-\lambda^2 t/2} \left(\mid \phi_\lambda(y) \mid^2 + \mid \phi_\lambda(z) \mid^2 \right) \right] \exp\left(\frac{-(u-v)^2}{4t} \right). \tag{4.2.21}$$

Moreover, since the kernel of $e^{-t\Delta_2}$ on the diagonal of $Y \times Y$ is bounded by $Ct^{-n/2}$, one finds that the kernels of $e^{-t\Delta_1}$ and $e^{-t\Delta_2}$ are exponentially small in t as $t \rightarrow 0^+$ for $u \neq v$, in that they are bounded by

$$C\, t^{-(n+1)/2} \exp\left(\frac{-(u-v)^2}{4t}\right),$$

where C is some constant, as $t \rightarrow 0^+$. Thus, the contribution outside the diagonal is asymptotically negligible, so that we are mainly interested in the contribution from the diagonal. For this purpose, we study $K(y,u;t)$, i.e. the kernel of $e^{-t\Delta_1} - e^{-t\Delta_2}$ at the point $(y,u;y,u)$ of $(Y \times \Re^+) \times (Y \times \Re^+)$. Defining $\text{sign}(\lambda) \equiv +1\ \forall \lambda \geq 0$, $\text{sign}(\lambda) \equiv -1\ \forall \lambda < 0$, one thus finds

$$\begin{aligned}
K(y,u;t) \equiv & \Big[W_A(y,u;y,u;t) - \widetilde{W}_B(y,u;y,u;t)\Big] \\
& + \Big[W_B(y,u;y,u;t) - \widetilde{W}_A(y,u;y,u;t)\Big] \\
= & \sum_{\lambda \geq 0} \mid \phi_\lambda(y) \mid^2 \left[\frac{e^{-\lambda^2 t}}{\sqrt{4\pi t}}\left(1 - e^{-u^2/t}\right) - \frac{e^{-\lambda^2 t}}{\sqrt{4\pi t}}\left(1 + e^{-u^2/t}\right)\right. \\
& \left. -(-\lambda) e^{2\lambda u} \text{erfc}\left(\frac{u}{\sqrt{t}} + \lambda\sqrt{t}\right)\right] \\
& + \sum_{\lambda < 0} \mid \phi_\lambda(y) \mid^2 \left[\frac{e^{-\lambda^2 t}}{\sqrt{4\pi t}}\left(1 + e^{-u^2/t}\right) + \lambda e^{-2\lambda u} \text{erfc}\left(\frac{u}{\sqrt{t}} - \lambda\sqrt{t}\right)\right. \\
& \left. - \frac{e^{-\lambda^2 t}}{\sqrt{4\pi t}}\left(1 - e^{-u^2/t}\right)\right] \\
= & \sum_{\lambda} \mid \phi_\lambda(y) \mid^2 \text{sign}(\lambda) \left[-\frac{e^{-\lambda^2 t} e^{-u^2/t}}{\sqrt{\pi t}}\right. \\
& \left. + \mid \lambda \mid e^{2|\lambda| u} \text{erfc}\left(\frac{u}{\sqrt{t}} + \mid \lambda \mid \sqrt{t}\right)\right] \\
= & \sum_{\lambda} \mid \phi_\lambda(y) \mid^2 \text{sign}(\lambda) \frac{\partial}{\partial u}\left[\frac{1}{2} e^{2|\lambda| u} \text{erfc}\left(\frac{u}{\sqrt{t}} + \mid \lambda \mid \sqrt{t}\right)\right].
\end{aligned} \quad (4.2.22)$$

Integration on $Y \times \Re^+$ and elementary rules for taking limits yield

$$K(t) \equiv \int_0^\infty \int_Y K(y,u;t)\, dy\, du = -\frac{1}{2} \sum_\lambda \text{sign}(\lambda)\ \text{erfc}(\mid \lambda \mid \sqrt{t}), \quad (4.2.23)$$

which implies (on differentiating the error function and using the signature of the eigenvalues)

$$K'(t) = \frac{1}{\sqrt{4\pi t}} \sum_{\lambda} \lambda e^{-\lambda^2 t}. \tag{4.2.24}$$

It is now necessary to derive the limiting behaviour of the integrated kernel (4.2.23) as $t \to \infty$ and as $t \to 0^+$. Indeed, denoting by h the degeneracy of the zero-eigenvalue, one finds

$$\lim_{t \to \infty} K(t) = -\frac{h}{2}, \tag{4.2.25}$$

whereas, as $t \to 0^+$, the following bound holds:

$$| K(t) | \leq \frac{1}{2} \sum_{\lambda} \mathrm{erfc}(| \lambda | \sqrt{t}) \leq \frac{1}{\sqrt{\pi}} \sum_{\lambda} e^{-\lambda^2 t} < C\, t^{-n/2}, \tag{4.2.26}$$

where C is a constant. Moreover, the result (4.2.25) may be supplemented by saying that $K(t) + \frac{h}{2}$ tends to 0 exponentially as $t \to \infty$. Thus, combining (4.2.25), (4.2.26) and this property, one finds that, for $\mathrm{Re}(s)$ sufficiently large, the integral

$$I(s) \equiv \int_0^\infty \left(K(t) + \frac{h}{2} \right) t^{s-1}\, dt, \tag{4.2.27}$$

converges. Integration by parts, definition of the Γ-function

$$\Gamma(z) \equiv \int_0^\infty t^{z-1} e^{-t}\, dt = k^z \int_0^\infty t^{z-1} e^{-kt}\, dt, \tag{4.2.28}$$

and careful consideration of positive and negative eigenvalues with their signatures then lead to

$$I(s) = -\frac{\Gamma\left(s + \frac{1}{2}\right)}{2s\sqrt{\pi}} \sum_{\lambda} \frac{\mathrm{sign}(\lambda)}{| \lambda |^{2s}} = -\frac{\Gamma\left(s + \frac{1}{2}\right)}{2s\sqrt{\pi}} \eta(2s). \tag{4.2.29}$$

The following analysis relies entirely on the *assumption* than an asymptotic expansion of the integrated kernel $K(t)$ exists as $t \to 0^+$. By writing such an expansion in the form

$$K(t) \sim \sum_{k \geq -m} a_k t^{k/2}, \tag{4.2.30}$$

Eqs. (4.2.27)–(4.2.30) yield (on splitting the integral (4.2.27) into an integral from 0 to 1 plus an integral from 1 to ∞)

$$\eta(2s) \sim -\frac{2s\sqrt{\pi}}{\Gamma\left(s + \frac{1}{2}\right)} \left[\frac{h}{2s} + \sum_{k=-m}^{N} \frac{a_k}{\left(\frac{k}{2} + s\right)} + \theta_N(s) \right]. \tag{4.2.31}$$

This is the analytic continuation of $\eta(2s)$ to the whole s-plane. Hence one finds

$$\eta(0) = -\Big(2a_0 + h\Big). \tag{4.2.32}$$

Regularity at the origin of the η-function is an important property in the theory of elliptic operators on manifolds. What one can prove (Atiyah *et al.* 1976, Gilkey 1995) is that the analytic continuation of the η-function of a given elliptic operator A:

$$\eta_A(s) \equiv \mathrm{Tr}\left[A \mid A \mid^{-s-1}\right], \tag{4.2.33}$$

is a meromorphic function which is regular at $s = 0$. Interestingly, such a property is stable under homotopy, i.e. under a smooth variation of A. More precisely, one considers a C^∞ one-parameter family A_u of elliptic operators, and one proves that the residue $R(A_u)$ at $s = 0$ of the corresponding η-functions is constant, i.e.

$$\frac{d}{du} R(A_u) = 0. \tag{4.2.34}$$

Thus, the general formula for the analytic continuation of the η-function of A reduces to (cf. Atiyah *et al.* 1976)

$$\eta(s) = \sum_{k=-m}^{N} \frac{b_k}{(s + \frac{k}{n})} + \phi_N(s) \ , \ \ k \neq 0, \tag{4.2.35}$$

where m is the dimension of the compact Riemannian manifold, n is the order of A, ϕ_N is holomorphic in the half-plane $\mathrm{Re}(s) > -\frac{N}{n}$, and is C^∞.

4.3 A further example

It is instructive to describe a simpler example of $\eta(0)$ calculation for first-order differential operators. For this purpose, we consider the first-order operator

$$A \equiv i \frac{d}{dx} + t, \tag{4.3.1}$$

where t is a real parameter lying in the open interval $]0, 1[$, and x is an angular coordinate on the circle. The boundary conditions on the eigenfunctions of A, denoted by f, are periodicity of period 2π:

$$f(x) = f(x + 2\pi). \tag{4.3.2}$$

Thus, since the eigenvalue equation for A:

$$i\frac{df}{dx} + tf = \lambda f \tag{4.3.3}$$

is solved by (f_0 being $f(x=0)$)

$$f = f_0 \, e^{-i(\lambda - t)x}, \tag{4.3.4}$$

one finds, by virtue of (4.3.2), the eigenvalue condition

$$e^{-i(\lambda - t)2\pi} = 1. \tag{4.3.5}$$

Equation (4.3.5) is solved by

$$\lambda = t \pm n \quad \forall n = 0, 1, 2, ... \tag{4.3.6}$$

One can now form the corresponding η-function:

$$\eta_t(s) = \sum_{r=0}^{\infty} (r+t)^{-s} - \sum_{r=1}^{\infty} (r-t)^{-s}, \tag{4.3.7}$$

which can be re-expressed as follows:

$$\eta_t(s) = t^{-s} + \sum_{r=1}^{\infty} \frac{\left[(r-t)^s - (r+t)^s\right]}{(r+t)^s (r-t)^s}$$

$$= t^{-s} - 2st \sum_{r=1}^{\infty} r^{-(s+1)} + s\Sigma_1. \tag{4.3.8}$$

In Eq. (4.3.8), Σ_1 is a series which is absolutely convergent in the neighbourhood of $s=0$. One thus finds

$$\eta_t(0) = 1 - 2t \lim_{s \to 0} s\zeta_R(s+1) = 1 - 2t, \tag{4.3.9}$$

where ζ_R is the Riemann ζ-function, defined as

$$\zeta_R(s) \equiv \sum_{l=1}^{\infty} l^{-s}. \tag{4.3.10}$$

The result (4.3.9) shows that, unless t takes the value $\frac{1}{2}$, there will be a 'spectral asymmetry' expressed by a non-vanishing value of $\eta_t(0)$.

It has been our choice to give a very elementary introduction to the subject of the η-function. The reader who is interested in advanced topics is referred to Berline *et al.* (1992), Branson and Gilkey (1992a,b), Booss-Bavnbek and Wojciechowski (1993), Grubb and Seeley (1995), Falomir *et al.* (1996a,b). In particular, the work of Grubb and Seeley (1995) has shown that, in general, the asymptotic expansion as $t \to 0^+$ of the integrated kernel (cf.

chapter 5) may contain logarithmic terms when Atiyah–Patodi–Singer boundary conditions are imposed (see Eq. (3.35) therein).

Further important work on the Dirac operator and its eigenvalues, and on spin-structures, can be found in Hortacsu *et al.* (1980), Atiyah and Singer (1984), Vafa and Witten (1984), Atiyah (1985), Dabrowski and Trautman (1986), Musto *et al.* (1986), Polychronakos (1987), Sitenko (1989), Bär (1992a–d), Mishchenko and Sitenko (1992,1993), Mostafazadeh (1994a,b), Connes (1995), Bär (1996b,c), Camporesi and Higuchi (1996), Kori (1996), Wojciechowski *et al.* (1996), Booss-Bavnbek *et al.* (1997), Carow-Watamura and Watamura (1997), Landi and Rovelli (1997), Mostafazadeh (1997), Vancea (1997).

4.4 Massless spin-1/2 fields

The quantum theory of fermionic fields can be expressed, following the ideas of Feynman, in terms of amplitudes of going from suitable fermionic data on a spacelike surface $\mathcal{S}_I$, say, to fermionic data on a spacelike surface $\mathcal{S}_F$. To make sure that the quantum boundary-value problem is well posed, one has actually to consider the Euclidean formulation, where the boundary three-surfaces, Σ_I and Σ_F, say, may be regarded as (compact) Riemannian three-manifolds bounding a Riemannian four-manifold. In the case of massless spin-1/2 fields, which are the object of our investigation, one thus deals with transition amplitudes

$$\mathcal{A}[\text{boundary data}] = \int e^{-I_E} \mathcal{D}\psi \, \mathcal{D}\widetilde{\psi}, \tag{4.4.1}$$

where I_E is the Euclidean action functional, and the integration is over all massless spin-1/2 fields matching the boundary data on Σ_I and Σ_F. The path-integral representation of the quantum amplitude (4.4.1) is then obtained with the help of Berezin integration rules, and one has a choice of non-local (D'Eath and Esposito 1991b) or local (D'Eath and Esposito 1991a) boundary conditions. The mathematical foundations of the former lie in the theory of spectral asymmetry and Riemannian geometry (Atiyah *et al.* 1975, 1976), and their formulation can be described as follows. In two-component spinor notation (see appendix 4.A), a massless spin-1/2 field in a four-manifold with positive-definite metric is represented by a pair $\left(\psi^A, \widetilde{\psi}_{A'}\right)$ of independent spinor fields, not related by any spinor conjugation. Suppose now that ψ^A and $\widetilde{\psi}^{A'}$ are expanded on a family of concentric three-spheres as

$$\psi^A = \frac{1}{2\pi}\tau^{-\frac{3}{2}} \sum_{n=0}^{\infty} \sum_{p,q=1}^{(n+1)(n+2)} \alpha_n^{pq} \Big[m_{np}(\tau)\rho^{nqA} + \widetilde{r}_{np}(\tau)\overline{\sigma}^{nqA} \Big], \tag{4.4.2}$$

$$\widetilde{\psi}^{A'} = \frac{1}{2\pi}\tau^{-\frac{3}{2}} \sum_{n=0}^{\infty} \sum_{p,q=1}^{(n+1)(n+2)} \alpha_n^{pq} \Big[\widetilde{m}_{np}(\tau)\overline{\rho}^{nqA'} + r_{np}(\tau)\sigma^{nqA'} \Big]. \tag{4.4.3}$$

With a standard notation, τ is the Euclidean-time coordinate which plays the role of a radial coordinate, and the block-diagonal matrices α_n^{pq} and the ρ- and σ-harmonics are described in detail in D'Eath and Halliwell (1987). One can now check that the harmonics ρ^{nqA} have positive eigenvalues for the intrinsic three-dimensional Dirac operator on S^3 (D'Eath and Halliwell 1987, Trautman 1992):

$$\mathcal{D}_{AB} \equiv {}_e n_{AB'} \, e_B^{\ \ B'j} \, {}^{(3)}D_j, \tag{4.4.4}$$

and similarly for the harmonics $\sigma^{nqA'}$ and the Dirac operator

$$\mathcal{D}_{A'B'} \equiv {}_e n_{BA'} \, e_{B'}^{\ \ Bj} \, {}^{(3)}D_j. \tag{4.4.5}$$

With our notation, ${}_e n_{AB'}$ is the Euclidean normal to the boundary, $e_B^{\ \ B'j}$ are the spatial components of the two-spinor version of the tetrad, and ${}^{(3)}D_j$ denotes three-dimensional covariant differentiation with respect to the Levi–Civita connection on S^3. By contrast, the harmonics $\overline{\sigma}^{nqA}$ and $\overline{\rho}^{nqA'}$ have negative eigenvalues for the operators (4.4.4) and (4.4.5), respectively.

The so-called *spectral* boundary conditions rely therefore on a non-local operation, i.e. the separation of the spectrum of a first-order elliptic operator (our (4.4.4) and (4.4.5)) into a positive and a negative part. They require that half of the spin-1/2 field should vanish on Σ_F, where this half is given by those modes $m_{np}(\tau)$ and $r_{np}(\tau)$ which multiply harmonics having positive eigenvalues for (4.4.4) and (4.4.5), respectively. The remaining half of the field should vanish on Σ_I, and is given by those modes $\widetilde{r}_{np}(\tau)$ and $\widetilde{m}_{np}(\tau)$ which multiply harmonics having negative eigenvalues for

(4.4.4) and (4.4.5), respectively. One thus writes

$$\left[\psi_{(+)}^{A}\right]_{\Sigma_F} = 0 \Longrightarrow \left[m_{np}\right]_{\Sigma_F} = 0, \tag{4.4.6}$$

$$\left[\widetilde{\psi}_{(+)}^{A'}\right]_{\Sigma_F} = 0 \Longrightarrow \left[r_{np}\right]_{\Sigma_F} = 0, \tag{4.4.7}$$

and

$$\left[\psi_{(-)}^{A}\right]_{\Sigma_I} = 0 \Longrightarrow \left[\widetilde{r}_{np}\right]_{\Sigma_I} = 0, \tag{4.4.8}$$

$$\left[\widetilde{\psi}_{(-)}^{A'}\right]_{\Sigma_I} = 0 \Longrightarrow \left[\widetilde{m}_{np}\right]_{\Sigma_I} = 0. \tag{4.4.9}$$

Massless spin-1/2 fields are studied here since they provide an interesting example of conformally invariant field theory for which the spectral boundary conditions (4.4.6)–(4.4.9) occur naturally already at the classical level (D'Eath and Halliwell 1987).

Section 4.5 is devoted to the evaluation of the $\zeta(0)$ value resulting from the boundary conditions (4.4.6)–(4.4.9). This yields the one-loop divergence of the quantum amplitude, and coincides with the conformal anomaly in our model. Essential self-adjointness is proved in section 4.6.

4.5 $\zeta(0)$ value with non-local boundary conditions

As shown in D'Eath and Esposito (1991a,b) and Esposito (1994a), the modes occurring in the expansions (4.4.2) and (4.4.3) obey a coupled set of equations, i.e.

$$\left(\frac{d}{d\tau} - \frac{\left(n+\frac{3}{2}\right)}{\tau}\right) x_{np} = E_{np}\, \widetilde{x}_{np}, \tag{4.5.1}$$

$$\left(-\frac{d}{d\tau} - \frac{\left(n+\frac{3}{2}\right)}{\tau}\right) \widetilde{x}_{np} = E_{np}\, x_{np}, \tag{4.5.2}$$

where x_{np} denotes m_{np} or r_{np}, and $\widetilde{x}_{np}$ denotes $\widetilde{m}_{np}$ or $\widetilde{r}_{np}$. Setting $E_{np} = M$ for simplicity of notation one thus finds, for all $n \geq 0$, the solutions of Eqs. (4.5.1) and (4.5.2) in the form

$$m_{np}(\tau) = \beta_{1,n}\sqrt{\tau} I_{n+1}(M\tau) + \beta_{2,n}\sqrt{\tau} K_{n+1}(M\tau), \tag{4.5.3}$$

$$r_{np}(\tau) = \beta_{1,n}\sqrt{\tau} I_{n+1}(M\tau) + \beta_{2,n}\sqrt{\tau} K_{n+1}(M\tau), \tag{4.5.4}$$

$$\widetilde{m}_{np}(\tau) = \beta_{1,n}\sqrt{\tau} I_{n+2}(M\tau) - \beta_{2,n}\sqrt{\tau} K_{n+2}(M\tau), \tag{4.5.5}$$

$$\widetilde{r}_{np}(\tau) = \beta_{1,n}\sqrt{\tau}I_{n+2}(M\tau) - \beta_{2,n}\sqrt{\tau}K_{n+2}(M\tau), \qquad (4.5.6)$$

where $\beta_{1,n}$ and $\beta_{2,n}$ are some constants. The insertion of (4.5.3)–(4.5.6) into the boundary conditions (4.4.6)–(4.4.9) leads to the equations (hereafter b and a are the radii of the two concentric three-sphere boundaries, with $b > a$, and we define $\beta_n \equiv \beta_{2,n}/\beta_{1,n}$)

$$I_{n+1}(Mb) + \beta_n K_{n+1}(Mb) = 0, \qquad (4.5.7)$$

for m_{np} and r_{np} modes, and

$$I_{n+2}(Ma) - \beta_n K_{n+2}(Ma) = 0, \qquad (4.5.8)$$

for $\widetilde{m}_{np}$ and $\widetilde{r}_{np}$ modes, with the same value of M. One thus finds two equivalent formulae for β_n:

$$\beta_n = -\frac{I_{n+1}(Mb)}{K_{n+1}(Mb)} = \frac{I_{n+2}(Ma)}{K_{n+2}(Ma)}, \qquad (4.5.9)$$

which lead to the eigenvalue condition

$$I_{n+1}(Mb)K_{n+2}(Ma) + I_{n+2}(Ma)K_{n+1}(Mb) = 0. \qquad (4.5.10)$$

The full degeneracy is $2(n+1)(n+2)$, for all $n \geq 0$, since each set of modes contributes to (4.5.7) and (4.5.8) with degeneracy $(n+1)(n+2)$ (D'Eath and Esposito 1991b).

We can now apply ζ-function regularization to evaluate the resulting conformal anomaly, following the algorithm developed in Barvinsky *et al.* (1992) and applied several times in the recent literature (Esposito *et al.* 1997). The basic properties are as follows. Let us denote by f_n the function occurring in the equation obeyed by the eigenvalues by virtue of the boundary conditions, after taking out fake roots (e.g. $x = 0$ is a fake root of order ν of the Bessel function $I_\nu(x)$). Let $d(n)$ be the degeneracy of the eigenvalues parametrized by the integer n. One can then define the function

$$I(M^2, s) \equiv \sum_{n=n_0}^{\infty} d(n) n^{-2s} \log f_n(M^2), \qquad (4.5.11)$$

and the work in Barvinsky *et al.* (1992) shows that such a function admits an analytic continuation to the complex-s plane as a meromorphic function with a simple pole at $s = 0$, in the form

$$\text{“}I(M^2, s)\text{”} = \frac{I_{\text{pole}}(M^2)}{s} + I^R(M^2) + \mathrm{O}(s). \qquad (4.5.12)$$

The function $I_{\rm pole}(M^2)$ is the residue at $s = 0$, and makes it possible to obtain the $\zeta(0)$ value as

$$\zeta(0) = I_{\log} + I_{\rm pole}(M^2 = \infty) - I_{\rm pole}(M^2 = 0), \qquad (4.5.13)$$

where $I_{\log}$ is the coefficient of the $\log(M)$ term in I^R as $M \rightarrow \infty$. The contributions $I_{\log}$ and $I_{\rm pole}(\infty)$ are obtained from the uniform asymptotic expansions of basis functions as $M \rightarrow \infty$ and their order $n \rightarrow \infty$ (appendix 4.B), while $I_{\rm pole}(0)$ is obtained by taking the $M \rightarrow 0$ limit of the eigenvalue condition, and then studying the asymptotics as $n \rightarrow \infty$. More precisely, $I_{\rm pole}(\infty)$ coincides with the coefficient of $\frac{1}{n}$ in the asymptotic expansion as $n \rightarrow \infty$ of

$$\frac{1}{2} d(n) \log\Big[\rho_\infty(n)\Big],$$

where $\rho_\infty(n)$ is the n-dependent term in the eigenvalue condition as $M \rightarrow \infty$ and $n \rightarrow \infty$. The $I_{\rm pole}(0)$ value is instead obtained as the coefficient of $\frac{1}{n}$ in the asymptotic expansion as $n \rightarrow \infty$ of

$$\frac{1}{2} d(n) \log\Big[\rho_0(n)\Big],$$

where $\rho_0(n)$ is the n-dependent term in the eigenvalue condition as $M \rightarrow 0$ and $n \rightarrow \infty$ (Barvinsky *et al.* 1992, Kamenshchik and Mishakov 1992).

In our problem, using the limiting form of Bessel functions when the argument tends to zero, one finds that the left-hand side of Eq. (4.5.10) is proportional to M^{-1} as $M \rightarrow 0$. Hence one has to multiply by M to get rid of fake roots. Moreover, in the uniform asymptotic expansion of Bessel functions as $M \rightarrow \infty$ and $n \rightarrow \infty$, both I and K functions contribute a $\frac{1}{\sqrt{M}}$ factor. These properties imply that $I_{\log}$ vanishes:

$$I_{\log} = \frac{1}{2} \sum_{l=1}^{\infty} 2l(l+1)\Big(1 - \frac{1}{2} - \frac{1}{2}\Big) = 0. \qquad (4.5.14)$$

Moreover,

$$I_{\rm pole}(\infty) = 0, \qquad (4.5.15)$$

since there is no n-dependent coefficient in the uniform asymptotic expansion of Eq. (4.5.10). Last, one finds

$$I_{\rm pole}(0) = 0, \qquad (4.5.16)$$

since the limiting form of Eq. (4.5.10) as $M \rightarrow 0$ and $n \rightarrow \infty$ is

$$\frac{2}{Ma}(b/a)^{n+1}.$$

The results (4.5.14)–(4.5.16), jointly with the general formula (4.5.13), lead to a vanishing value of the one-loop divergence (Esposito 1997a):

$$\zeta(0) = 0. \tag{4.5.17}$$

Our detailed calculation shows that, in flat Euclidean four-space, the conformal anomaly for a massless spin-1/2 field subject to non-local boundary conditions of the spectral type on two concentric three-spheres vanishes, as in the case when the same field is subject to the local boundary conditions

$$\sqrt{2}\; {}_e n_A^{\ A'} \; \psi^A = \pm \widetilde{\psi}^{A'} \quad \text{on } \Sigma_I \text{ and } \Sigma_F. \tag{4.5.18}$$

If Eq. (4.5.18) holds and the spin-1/2 field is massless, the work in Kamenshchik and Mishakov (1994) shows in fact that $\zeta(0) = 0$.

Backgrounds given by flat Euclidean four-space bounded by two concentric three-spheres are not the ones occurring in the Hartle–Hawking proposal for quantum cosmology, where the initial three-surface Σ_I shrinks to a point (Hartle and Hawking 1983, Hawking 1984). Nevertheless, they are relevant for the quantization programme of gauge fields and gravitation in the presence of boundaries (Esposito *et al.* 1997). In particular, similar techniques have been used in section 5 of Esposito and Kamenshchik (1996) to study a two-boundary problem for simple supergravity subject to spectral boundary conditions in the axial gauge. One then finds the eigenvalue condition

$$I_{n+2}(Mb)K_{n+3}(Ma) + I_{n+3}(Ma)K_{n+2}(Mb) = 0, \tag{4.5.19}$$

for all $n \geq 0$. The analysis of Eq. (4.5.19) along the same lines of what we have done for Eq. (4.5.10) shows that three-dimensional transverse-traceless gravitino modes yield a vanishing contribution to $\zeta(0)$, unlike three-dimensional transverse-traceless modes for gravitons, which instead contribute -5 to $\zeta(0)$ (Esposito *et al.* 1994b).

Thus, the results in Esposito and Kamenshchik (1996) seem to show that, at least in finite regions bounded by one three-sphere or two concentric three-spheres, simple supergravity is not one-loop finite in the presence of boundaries. Of course, more work is in order to check this property, and then compare it with the finiteness of scattering problems suggested by D'Eath (1996). Further progress is thus likely to occur by virtue of the fertile interplay of geometric and analytic techniques in the investigation of heat-kernel asymptotics and (one-loop) quantum cosmology.

4.6 Essential self-adjointness

This section is devoted to the mathematical foundations of the one-loop analysis described so far. Indeed, a naturally occurring question is whether a *unique* self-adjoint extension for the spin-1/2 boundary-value problem exists when the boundary conditions (4.4.6)–(4.4.9) are imposed, since otherwise different self-adjoint extensions would lead to different spectra and hence the trace anomaly would be ill defined. We shall see that this is not the case, and hence the boundary conditions are enough to determine a unique, real and positive spectrum for the squared Dirac operator (from which the $\zeta(0)$ value can be evaluated as we just did).

We here consider the portion of flat Euclidean four-space which is bounded by a three-sphere of radius a. On inserting the expansions (4.4.2) and (4.4.3) into the massless spin-1/2 action

$$I_E = \frac{i}{2} \int_M \left[\tilde{\psi}^{A'} \left(\nabla_{AA'} \psi^A \right) - \left(\nabla_{AA'} \tilde{\psi}^{A'} \right) \psi^A \right] \sqrt{\det g} \, d^4x, \quad (4.6.1)$$

and studying the spin-1/2 eigenvalue equations, one finds that the modes obey the second-order differential equations (cf. (4.5.1) and (4.5.2))

$$P_n \widetilde{m}_{np} = P_n \widetilde{m}_{n,p+1} = P_n \widetilde{r}_{np} = P_n \widetilde{r}_{n,p+1} = 0, \quad (4.6.2)$$

$$Q_n r_{np} = Q_n r_{n,p+1} = Q_n m_{np} = Q_n m_{n,p+1} = 0, \quad (4.6.3)$$

where

$$P_n \equiv \frac{d^2}{d\tau^2} + \left[E_n^2 - \frac{((n+2)^2 - \frac{1}{4})}{\tau^2} \right], \quad (4.6.4)$$

$$Q_n \equiv \frac{d^2}{d\tau^2} + \left[E_n^2 - \frac{((n+1)^2 - \frac{1}{4})}{\tau^2} \right], \quad (4.6.5)$$

and E_n are the eigenvalues of the mode-by-mode form of the Dirac operator (Esposito 1994a).

The spectral boundary conditions (4.4.6)–(4.4.9) imply that, in our case,

$$m_{np}(a) = 0, \tag{4.6.6}$$

$$r_{np}(a) = 0. \tag{4.6.7}$$

Thus, one studies the one-dimensional operators Q_n defined in (4.6.5), and the eigenmodes are requested to be regular at the origin, and to obey (4.6.6) and (4.6.7) on S^3. For this purpose, it is convenient to consider, for all $n \geq 0$, the differential operators

$$\widetilde{Q}_n \equiv -\frac{d^2}{d\tau^2} + \frac{((n+1)^2 - \frac{1}{4})}{\tau^2}. \tag{4.6.8}$$

These are particular cases of a large class of operators considered in the literature. They can be studied by using the following definitions and theorems from Reed and Simon (1975) (cf. Weidmann 1980).

Definition 4.6.1 The function V is in the *limit circle* case at zero if for some, and therefore all λ, *all* solutions of the equation

$$-\varphi''(x) + V(x)\varphi(x) = \lambda\varphi(x) \tag{4.6.9}$$

are square integrable at zero.

Definition 4.6.2 If $V(x)$ is not in the limit circle case at zero, it is said to be in the *limit point* case at zero.

Theorem 4.6.1 (Weyl's limit point – limit circle criterion) Let V be a continuous real-valued function on $(0, \infty)$. Then $\mathcal{O} \equiv -\frac{d^2}{dx^2} + V(x)$ is essentially self-adjoint on $C_0^\infty(0, \infty)$ if and only if $V(x)$ is in the limit point case at both zero and infinity.

Theorem 4.6.2 Let V be continuous and *positive* near zero. If $V(x) \geq \frac{3}{4}x^{-2}$ near zero, then $\mathcal{O}$ is in the limit point case at zero.

Theorem 4.6.3 Let V be differentiable on $]0, \infty[$ and bounded above by K on $[1, \infty[$. Suppose that

(i) $\int_1^\infty \frac{dx}{\sqrt{K - V(x)}} = \infty$,

(ii) $V'(x) \mid V(x) \mid^{-\frac{3}{2}}$ is bounded near infinity.

Then $V(x)$ is in the limit point case at ∞.

In other words, a necessary and sufficient condition for the existence of a unique self-adjoint extension of $\mathcal{O}$, is that its eigenfunctions should fail to be square integrable at zero and at infinity. We will not give many technical details to prove *all* these theorems, since we are more interested in the applications of the general theory. However, we find it helpful for the general reader to point out that, when $V(x)$ takes the form $\frac{c}{x^2}$ for $c > 0$, theorem 4.6.2 can be proved as follows (Reed and Simon 1975). The equation

$$-\varphi''(x) + \frac{c}{x^2}\varphi(x) = 0 \tag{4.6.10}$$

admits solutions of the form x^α, where α takes the values

$$\alpha_1 = \frac{1}{2} + \frac{1}{2}\sqrt{1+4c}, \tag{4.6.11}$$

$$\alpha_2 = \frac{1}{2} - \frac{1}{2}\sqrt{1+4c}. \tag{4.6.12}$$

When $\alpha = \alpha_1$ the solution is obviously square integrable at zero. However, when $\alpha = \alpha_2$, the solution is square integrable at zero if and only if $\alpha_2 > -\frac{1}{2}$, which implies $c < \frac{3}{4}$. By virtue of the definitions 4.6.1 and 4.6.2, this means that $V(x)$ is in the limit point at zero if and only if $c \geq \frac{3}{4}$. However, in our problem we are interested in the closed interval $[0, a]$, where a is finite. Thus, the previous theorems can only be used after relating the original problem to another one involving the infinite interval $(0, \infty)$. For this purpose, we perform a double change of variables. The first one affects the independent variable, whereas the second one is performed on the dependent variable so that the operator takes the form (4.6.9) with the appropriate (semi)infinite domain (Esposito *et al.* 1996).

Given Eq. (4.6.10) with $x \in [0, a]$ (see (4.6.9)) we perform the change

$$x \equiv a\left(1 - e^{-y}\right) \Rightarrow y \in (0, \infty). \tag{4.6.13}$$

This yields the equation

$$\ddot{\varphi}(y) + \dot{\varphi}(y) - \frac{c(n)}{(e^y - 1)^2}\varphi(y) = 0 \quad y \in (0, \infty), \tag{4.6.14}$$

where $c(n) \equiv (n+1)^2 - \frac{1}{4}$ and the *dot* denotes differentiation with respect to y. The further change

$$\varphi(y) \equiv e^{-\frac{y}{2}}\chi(y) \tag{4.6.15}$$

leads to

$$-\ddot{\chi}(y) + \left[\frac{1}{4} + \frac{c(n)}{(e^y - 1)^2}\right] \chi = 0 \quad y \in (0, \infty), \tag{4.6.16}$$

which has the structure of the left-hand side of Eq. (4.6.9) with a suitable domain, so that Weyl's theorem is directly applicable. Indeed, when $y \to \infty$, the conditions (i) and (ii) of theorem 4.6.3 are clearly satisfied by the potential in Eq. (4.6.16), and hence such a $V(y)$ is in the limit point at ∞ (in that case, the constant of theorem 4.6.3 is $K(n) \equiv \frac{1}{4} + \frac{c(n)}{(e-1)^2}$). Moreover, when $y \to 0$, $V(y)$ tends to $\frac{1}{4} + \frac{c(n)}{y^2} \geq \frac{1}{4} + \frac{3}{4}\frac{1}{y^2}$, since $n \geq 0$ in (4.6.8). Thus, the condition of theorem 4.6.2 is satisfied and we are in the limit point at zero as well. By virtue of theorem 4.6.1, these properties imply that our operators (4.6.8) are related, through (4.6.13) and (4.6.15), to the second-order elliptic operators

$$T_n \equiv -\frac{d^2}{dy^2} + \frac{1}{4} + \frac{((n+1)^2 - \frac{1}{4})}{(e^y - 1)^2}, \tag{4.6.17}$$

which are essentially self-adjoint on the space of C^∞ functions on $(0, \infty)$ with compact support.

The mathematical literature on self-adjointness properties of the Dirac operator in the pseudo-Riemannian case is indeed quite rich (see, for example, Chernoff (1977) and references therein). However, one-loop quantum cosmology needs a rigorous proof of essential self-adjointness of *elliptic* operators with local or non-local boundary conditions. The calculation of conformal anomalies for massless spin-1/2 fields on flat Euclidean manifolds with boundary, initiated in D'Eath and Esposito (1991a,b) and confirmed in Kamenshchik and Mishakov (1992,1993), has led over the past few years to a substantial improvement of the understanding of one-loop divergences in quantum field theory. Various independent techniques, which often rely on the structures applied in the spin-1/2 analysis, have been used to deal with the whole set of perturbative modes for Euclidean Maxwell theory and Euclidean quantum gravity (Esposito 1994a,b, Esposito *et al.* 1994a,b, 1995a,b, 1997). Nevertheless, a systematic analysis aimed at explaining why the various $\zeta(0)$ values are well defined was still lacking in the literature.

Our investigation, which relies on the work by Esposito *et al.*

(1996), represents the first step in this direction, in the case of the squared Dirac operator with the spectral boundary conditions (4.4.6)–(4.4.9). Note that such non-local boundary conditions play a crucial role in our proof. As shown on page 153 of Reed and Simon (1975), to prove theorem 4.6.1 one has to choose at some stage a point $c \in (0, \infty)$ and then consider the operator

$$A \equiv -\frac{d^2}{dx^2} + V(x)$$

on the domain

$$D(A) \equiv \{\omega : \omega \in C^\infty(0, c), \omega = 0 \text{ near zero}, \omega(c) = 0\}.$$

As far as we can see, this scheme is only compatible with the boundary conditions (4.4.6)–(4.4.9). Thus, we should stress that the extension to the local boundary conditions studied in D'Eath and Esposito (1991a) (see section 6.3) remains the main open problem in our investigation. For this purpose, the calculation of deficiency indices (Reed and Simon 1975) appears more appropriate. This would complete the analysis in Esposito (1994a, 1995), where the existence of self-adjoint extensions is proved by showing that a linear, anti-involutory operator F exists which is norm-preserving, commutes with the Dirac operator $\mathcal{D}$ and maps the domain of $\mathcal{D}$ into itself.

Moreover, it remains to be seen how to apply similar techniques to the analysis of higher-spin fields. These are gauge fields and gravitation, which obey a complicated set of mixed boundary conditions (Esposito *et al.* 1997). The mixed nature of the boundary conditions results from the request of their invariance under infinitesimal gauge transformations (Avramidi and Esposito 1997c). In particular, for the gravitational field, at least five different sets of mixed boundary conditions have been proposed in the literature (Barvinsky 1987, Luckock and Moss 1989, Luckock 1991, Esposito and Kamenshchik 1995, Marachevsky and Vassilevich 1996, Moss and Silva 1997). If this investigation could be completed, it would put on solid ground the current work on trace anomalies and one-loop divergences on manifolds with boundary, and it would add evidence in favour of quantum cosmology being at the very heart of many exciting developments in quantum field theory, analysis and differential geometry (section 6.6).

Appendix 4.A

Two-component spinor calculus is an elegant and powerful tool for studying some properties of classical field theories in four-dimensional space-time models. Within this framework, the basic object is *spin-space*, a two-dimensional complex vector space S with a symplectic form ε, i.e. an anti-symmetric complex bilinear form. Unprimed spinor indices $A, B, \ldots$ take the values $0, 1$, whereas primed spinor indices $A', B', \ldots$ take the values $0', 1'$ since there are actually two such spaces: unprimed spin-space (S, ε) and primed spin-space (S', ε'). The whole two-spinor calculus in *Lorentzian* four-manifolds relies on three fundamental properties (Penrose 1960, Penrose and Rindler 1984,1986, Ward and Wells 1990).

(i) The isomorphism between $\left(S, \varepsilon_{AB}\right)$ and its dual $\left(S^*, \varepsilon^{AB}\right)$. This is provided by the symplectic form ε, which raises and lowers indices according to the rules

$$\varepsilon^{AB} \; \varphi_B = \varphi^A \; \in \; S, \tag{4.A.1}$$

$$\varphi^B \; \varepsilon_{BA} = \varphi_A \; \in \; S^*. \tag{4.A.2}$$

Thus, since

$$\varepsilon_{AB} = \varepsilon^{AB} = \begin{pmatrix} 0 & 1 \\ -1 & 0 \end{pmatrix}, \tag{4.A.3}$$

one finds in components $\varphi^0 = \varphi_1, \varphi^1 = -\varphi_0$.

Similarly, one has the isomorphism $\left(S', \varepsilon_{A'B'}\right) \cong \left((S')^*, \varepsilon^{A'B'}\right)$, which implies

$$\varepsilon^{A'B'} \; \varphi_{B'} = \varphi^{A'} \; \in \; S', \tag{4.A.4}$$

$$\varphi^{B'} \; \varepsilon_{B'A'} = \varphi_{A'} \; \in \; (S')^*, \tag{4.A.5}$$

where

$$\varepsilon_{A'B'} = \varepsilon^{A'B'} = \begin{pmatrix} 0' & 1' \\ -1' & 0' \end{pmatrix}. \tag{4.A.6}$$

(ii) The (anti-)isomorphism between $\left(S, \varepsilon_{AB}\right)$ and $\left(S', \varepsilon_{A'B'}\right)$, called *complex conjugation*, and denoted by an overbar. According to a standard convention, one has

$$\overline{\psi^A} \equiv \overline{\psi}^{A'} \; \in \; S', \tag{4.A.7}$$

$$\overline{\psi^{A'}} \equiv \overline{\psi}^{A} \in S. \tag{4.A.8}$$

Thus, complex conjugation maps elements of a spin-space to elements of the *complementary* spin-space. Hence some authors say that it is an anti-isomorphism (Stewart 1991). In components, if w^A is viewed as $w^A = \begin{pmatrix} \alpha \\ \beta \end{pmatrix}$, the action of (4.A.7) leads to

$$\overline{w^A} \equiv \overline{w}^{A'} \equiv \begin{pmatrix} \overline{\alpha} \\ \overline{\beta} \end{pmatrix}, \tag{4.A.9}$$

whereas, if $z^{A'} = \begin{pmatrix} \gamma \\ \delta \end{pmatrix}$, then (4.A.8) leads to

$$\overline{z^{A'}} \equiv \overline{z}^{A} = \begin{pmatrix} \overline{\gamma} \\ \overline{\delta} \end{pmatrix}. \tag{4.A.10}$$

With our notation, $\overline{\alpha}$ denotes complex conjugation of the function α, and so on. Note that the symplectic structure is preserved by complex conjugation, since $\overline{\varepsilon}_{A'B'} = \varepsilon_{A'B'}$.

(iii) The isomorphism between the tangent space T at a point of space-time and the tensor product of the unprimed spin-space $\left(S, \varepsilon_{AB}\right)$ and the primed spin-space $\left(S', \varepsilon_{A'B'}\right)$:

$$T \cong \left(S, \varepsilon_{AB}\right) \otimes \left(S', \varepsilon_{A'B'}\right). \tag{4.A.11}$$

The Infeld–van der Waerden symbols $\sigma^a{}_{AA'}$ and $\sigma_a{}^{AA'}$ express this isomorphism, and the correspondence between a vector v^a and a spinor $v^{AA'}$ is given by

$$v^{AA'} \equiv v^a \, \sigma_a{}^{AA'}, \tag{4.A.12}$$

$$v^a \equiv v^{AA'} \, \sigma^a{}_{AA'}. \tag{4.A.13}$$

These mixed spinor-tensor symbols obey the identities (Ward and Wells 1990)

$$\overline{\sigma}_a{}^{AA'} = \sigma_a{}^{AA'}, \tag{4.A.14}$$

$$\sigma_a{}^{AA'} \, \sigma^b{}_{AA'} = \delta_a{}^b, \tag{4.A.15}$$

$$\sigma_a{}^{AA'} \, \sigma^a{}_{BB'} = \varepsilon_B{}^A \, \varepsilon_{B'}{}^{A'}, \tag{4.A.16}$$

$$\sigma_{[a}{}^{AA'} \, \sigma_{b]A}{}^{B'} = -\frac{i}{2} \, \varepsilon_{abcd} \, \sigma^{cAA'} \, \sigma^d{}_A{}^{B'}. \tag{4.A.17}$$

Similarly, a one-form ω_a has a spinor equivalent

$$\omega_{AA'} \equiv \omega_a \, \sigma^a{}_{AA'}, \tag{4.A.18}$$

whereas the spinor equivalent of the metric is

$$\eta_{ab}\, \sigma^{a}{}_{AA'}\, \sigma^{b}{}_{BB'} \equiv \varepsilon_{AB}\, \varepsilon_{A'B'}. \tag{4.A.19}$$

In particular, in Minkowski space-time, Eqs. (4.A.12) and (4.A.17) enable one to express a coordinate system in 2×2 matrix form

$$x^{AA'} = \frac{1}{\sqrt{2}} \begin{pmatrix} x^0 + x^3 & x^1 - ix^2 \\ x^1 + ix^2 & x^0 - x^3 \end{pmatrix}. \tag{4.A.20}$$

More precisely, in a curved space-time, one should write the following equation to obtain the spinor equivalent of a vector:

$$u^{AA'} = u^{a}\, e_{a}^{\ \hat{c}}\, \sigma_{\hat{c}}^{\ AA'}, \tag{4.A.21}$$

where $e_{a}^{\ \hat{c}}$ is a standard notation for the tetrad, and

$$e_{a}^{\ \hat{c}}\, \sigma_{\hat{c}}^{\ AA'} \equiv e_{a}^{\ AA'} \tag{4.A.22}$$

is called the *soldering form* (Ashtekar 1988, 1991). This is, by construction, a spinor-valued one-form, which encodes the relevant information about the metric g (and hence about the curvature), because (see (1.2.2))

$$g_{ab} = e_{a}^{\ \hat{c}}\, e_{b}^{\ \hat{d}}\, \eta_{\hat{c}\hat{d}}, \tag{4.A.23}$$

η being the Minkowskian metric of the so-called 'internal space'. The condition for a connection to be metric-compatible,

$$\nabla g = 0, \tag{4.A.24a}$$

is re-expressed by a stronger condition on the soldering form, i.e.

$$\nabla_{a}\, e_{b}^{\ CF'} = 0. \tag{4.A.24b}$$

By virtue of (4.A.23), the equation (4.A.24*b*) is a sufficient condition for the validity of (4.A.24*a*), but is not implied from it.

In the Lorentzian-signature case, the Maxwell two-form $F \equiv F_{ab} dx^{a} \wedge dx^{b}$ can be written spinorially (Ward and Wells 1990) as

$$F_{AA'BB'} = \frac{1}{2}\Big(F_{AA'BB'} - F_{BB'AA'}\Big) = \varphi_{AB}\, \varepsilon_{A'B'} + \varphi_{A'B'}\, \varepsilon_{AB}, \tag{4.A.25}$$

where

$$\varphi_{AB} \equiv \frac{1}{2} F_{AC'B}^{\quad\ \ C'} = \varphi_{(AB)}, \tag{4.A.26}$$

$$\varphi_{A'B'} \equiv \frac{1}{2} F_{CB'}^{\quad\ C}{}_{A'} = \varphi_{(A'B')}. \tag{4.A.27}$$

These formulae are obtained by applying the identity

$$T_{AB} - T_{BA} = \varepsilon_{AB}\, T_{C}^{\ \ C} \tag{4.A.28}$$

to express $\frac{1}{2}\Big(F_{AA'BB'} - F_{AB'BA'}\Big)$ and $\frac{1}{2}\Big(F_{AB'BA'} - F_{BB'AA'}\Big)$. Note also that round brackets (AB) denote (as usual) symmetrization over the spinor indices A and B, and that the anti-symmetric part of φ_{AB} vanishes by virtue of the anti-symmetry of F_{ab}, since (Ward and Wells 1990) $\varphi_{[AB]} = \frac{1}{4}\varepsilon_{AB}\, F_{CC'}{}^{CC'} = \frac{1}{2}\varepsilon_{AB}\, \eta^{cd}\, F_{cd} = 0$. Last, but not least, in the Lorentzian case

$$\overline{\varphi_{AB}} \equiv \overline{\varphi}_{A'B'} = \varphi_{A'B'}. \tag{4.A.29}$$

The symmetric spinor fields φ_{AB} and $\varphi_{A'B'}$ are the anti-self-dual and self-dual parts of the gauge curvature two-form, respectively.

Similarly, the Weyl curvature $C^a{}_{bcd}$, i.e. the part of the Riemann curvature tensor invariant under conformal rescalings of the metric, may be expressed spinorially, omitting soldering forms for simplicity of notation, as

$$C_{abcd} = \psi_{ABCD}\, \varepsilon_{A'B'}\, \varepsilon_{C'D'} + \overline{\psi}_{A'B'C'D'}\, \varepsilon_{AB}\, \varepsilon_{CD}. \tag{4.A.30}$$

It should be emphasized that it is exactly the omission of soldering forms that makes it possible to achieve the most transparent translation of tensor fields into their spinor equivalent (although we acknowledge that other authors might want to express the opposite view).

In canonical gravity, two-component spinors lead to a considerable simplification of calculations. Denoting by n^μ the future-pointing unit timelike normal to a spacelike three-surface, its spinor version obeys the relations

$$n_{AA'}\, e^{AA'}{}_i = 0, \tag{4.A.31}$$

$$n_{AA'}\, n^{AA'} = 1. \tag{4.A.32}$$

Denoting by h the induced metric on the three-surface, other useful relations are (D'Eath 1984, Esposito 1994a)

$$h_{ij} = -e_{AA'i}\, e^{AA'}{}_j, \tag{4.A.33}$$

$$e^{AA'}{}_0 = N\, n^{AA'} + N^i\, e^{AA'}{}_i, \tag{4.A.34}$$

$$n_{AA'}\, n^{BA'} = \frac{1}{2}\varepsilon_A{}^B, \tag{4.A.35}$$

$$n_{AA'}\, n^{AB'} = \frac{1}{2}\varepsilon_{A'}{}^{B'}, \tag{4.A.36}$$

$$n_{[EB'}\, n_{A]A'} = \frac{1}{4}\varepsilon_{EA}\, \varepsilon_{B'A'}, \tag{4.A.37}$$

$$e_{AA'j}\, e^{AB'}{}_{k} = -\frac{1}{2} h_{jk}\, \varepsilon_{A'}{}^{B'} - i\varepsilon_{jkl}\sqrt{\det h}\, n_{AA'}\, e^{AB'l}. \quad (4.A.38)$$

In Eq. (4.A.34), N and N^i are the lapse and shift functions respectively (Esposito 1994a).

To obtain the space-time curvature, we first need to define the spinor covariant derivative $\nabla_{AA'}$. If θ, ϕ, ψ are spinor fields, $\nabla_{AA'}$ is a map such that (Penrose and Rindler 1984, Stewart 1991)

(1) $\nabla_{AA'}(\theta + \phi) = \nabla_{AA'}\theta + \nabla_{AA'}\phi$ (i.e. linearity);

(2) $\nabla_{AA'}(\theta\psi) = \left(\nabla_{AA'}\theta\right)\psi + \theta\left(\nabla_{AA'}\psi\right)$ (i.e. Leibniz rule);

(3) $\psi = \nabla_{AA'}\theta$ implies $\overline{\psi} = \nabla_{AA'}\overline{\theta}$ (i.e. reality condition);

(4) $\nabla_{AA'}\varepsilon_{BC} = \nabla_{AA'}\varepsilon^{BC} = 0$, i.e. the symplectic form may be used to raise or lower indices within spinor expressions acted upon by $\nabla_{AA'}$, in addition to the usual metricity condition $\nabla g = 0$, which involves instead the product of two ε-symbols;

(5) $\nabla_{AA'}$ commutes with any index substitution not involving A, A';

(6) for any function f, one finds $\left(\nabla_a\nabla_b - \nabla_b\nabla_a\right)f = 2S_{ab}{}^{c}\, \nabla_c f$, where $S_{ab}{}^{c}$ is the torsion tensor;

(7) for any derivation D acting on spinor fields, a spinor field $\xi^{AA'}$ exists such that $D\psi = \xi^{AA'}\, \nabla_{AA'}\psi, \forall\psi$.

As proved in Penrose and Rindler (1984), such a spinor covariant derivative exists and is unique.

If Lorentzian space-time is replaced by a complex or real Riemannian four-manifold, an important modification of the formalism should be made, since the (anti-)isomorphism between unprimed and primed spin-space no longer exists. This means that primed spinors can no longer be regarded as complex conjugates of unprimed spinors, or viceversa, as in (4.A.7) and (4.A.8). In particular, Eqs. (4.A.25) and (4.A.30) should be re-written as

$$F_{AA'BB'} = \varphi_{AB}\, \varepsilon_{A'B'} + \widetilde{\varphi}_{A'B'}\, \varepsilon_{AB}, \quad (4.A.39)$$

$$C_{abcd} = \psi_{ABCD}\, \varepsilon_{A'B'}\, \varepsilon_{C'D'} + \widetilde{\psi}_{A'B'C'D'}\, \varepsilon_{AB}\, \varepsilon_{CD}. \quad (4.A.40)$$

With our notation, $\varphi_{AB}, \widetilde{\varphi}_{A'B'}$, as well as $\psi_{ABCD}, \widetilde{\psi}_{A'B'C'D'}$ are *completely independent* symmetric spinor fields, not related by any conjugation.

Indeed, a conjugation can still be defined in the real Riemannian case, but it no longer relates $\left(S, \varepsilon_{AB}\right)$ to $\left(S', \varepsilon_{A'B'}\right)$. It is instead an anti-involutory operation which maps elements of a spin-space (either unprimed or primed) to elements of the *same* spin-space. By anti-involutory we mean that, when applied twice to a spinor with an odd number of indices, it yields the same spinor with the opposite sign, i.e. its square is minus the identity, whereas the square of complex conjugation as defined in (4.A.9) and (4.A.10) equals the identity. Following Woodhouse (1985) and Esposito (1994a), *Euclidean conjugation*, denoted by a *dagger*, is defined as follows:

$$\left(w^A\right)^\dagger \equiv \begin{pmatrix} \overline{\beta} \\ -\overline{\alpha} \end{pmatrix}, \tag{4.A.41}$$

$$\left(z^{A'}\right)^\dagger \equiv \begin{pmatrix} -\overline{\delta} \\ \overline{\gamma} \end{pmatrix}. \tag{4.A.42}$$

This means that, in flat Euclidean four-space, a unit 2×2 matrix $\delta_{BA'}$ exists such that

$$\left(w^A\right)^\dagger \equiv \varepsilon^{AB} \, \delta_{BA'} \, \overline{w}^{A'}. \tag{4.A.43}$$

Bearing in mind what we know from section 3.6, we can say that we obtain a complex structure (our Euclidean conjugation) by composition of a symplectic and an Euclidean structure (our ε^{AB} and $\delta_{BA'}$, respectively). We are here using the freedom to regard w^A either as an $SL(2,C)$ spinor for which complex conjugation can be defined, or as an $SU(2)$ spinor for which Euclidean conjugation is instead available. The soldering forms for $SU(2)$ spinors only involve spinor indices of the same spin-space, i.e. $\widetilde{e}_i^{\ AB}$ and $\widetilde{e}_i^{\ A'B'}$ (cf. Ashtekar 1991). More precisely, denoting by E_a^i a real *triad*, where $i = 1,2,3$, and by $\tau^a{}_A{}^B$ the three Pauli matrices obeying the identity

$$\tau^a{}_A{}^B \, \tau^b{}_B{}^D = i \, \varepsilon^{abc} \, \tau_{cA}{}^D + \delta^{ab} \, \delta_A{}^D, \tag{4.A.44}$$

the $SU(2)$ soldering forms are defined by

$$\widetilde{e}^j{}_A{}^B \equiv -\frac{i}{\sqrt{2}} \, E_a^j \, \tau^a{}_A{}^B. \tag{4.A.45}$$

Note that our conventions differ from those in Ashtekar (1991), i.e. we use $\tilde{e}$ instead of σ, and a, b for Pauli-matrix indices, i, j for tangent-space indices on a three-manifold Σ, to agree with our previous notation. The soldering form in (4.A.45) provides an isomorphism between the three-real-dimensional tangent space at each point of Σ, and the three-real-dimensional vector space of 2×2 trace-free Hermitian matrices. The Riemannian three-metric on Σ is then given by

$$h^{ij} = -\tilde{e}^{i}{}_{A}{}^{B} \, \tilde{e}^{j}{}_{B}{}^{A}. \tag{4.A.46}$$

Appendix 4.B

The uniform asymptotic expansions of Bessel functions of large order are frequently applied in the literature on ζ-function methods and one-loop divergences, but only a limited number of research workers seem to be aware of how this hard piece of asymptotic analysis is actually derived. In this appendix, relying on the seminal paper by Olver (1954), we describe in detail the whole construction.

The starting point is the remark that the functions $\sqrt{z} I_n(nz)$ and $\sqrt{z} K_n(nz)$ satisfy the differential equation

$$\frac{d^2 w}{dz^2} = \left[\left(1 + \frac{1}{z^2} \right) n^2 - \frac{1}{4z^2} \right] w. \tag{4.B.1}$$

The transition points of Eq. (4.B.1) are given by 0 and $\pm i$. The analysis is confined to the half-plane $| \arg z | < \frac{\pi}{2}$. Nothing is lost by doing so, because the results may be extended to other phase ranges, not including the imaginary axis, by means of the continuation formulae

$$I_\nu(z e^{m\pi i}) = e^{m\nu\pi i} I_\nu(z), \tag{4.B.2}$$

$$K_\nu(z e^{m\pi i}) = e^{-m\nu\pi i} K_\nu(z) - \pi i \frac{\sin(m\nu\pi)}{\sin(\nu\pi)} I_\nu(z), \tag{4.B.3}$$

where m is an arbitrary integer. It is now convenient to change both the independent and the dependent variable in Eq. (4.B.1), with the help of the differential equations

$$\left(\frac{dy}{dz} \right)^2 = 1 + \frac{1}{z^2}, \tag{4.B.4}$$

$$W = \left(\frac{dy}{dz} \right)^{1/2} w = \left(1 + \frac{1}{z^2} \right)^{1/4} w. \tag{4.B.5}$$

The attention is then focused on the differential equation for W,

$$\frac{d^2W}{dy^2} = (n^2 + f(y))W, \tag{4.B.6}$$

where, setting $\dot{z} \equiv \frac{dz}{dy}$, one finds

$$f(y) \equiv -\frac{\dot{z}^2}{4z^2} + \dot{z}^{1/2}\frac{d^2}{dy^2}(\dot{z}^{-1/2}). \tag{4.B.7}$$

Note now that Eq. (4.B.4) may be integrated to give

$$y(z) = \sqrt{1+z^2} + \log\frac{z}{1+\sqrt{1+z^2}}, \tag{4.B.8}$$

where the integration constant has been set to zero for convenience. On taking y as independent variable, the half-plane

$$|\arg z| < \frac{\pi}{2}$$

is mapped on the domain, say D, containing the half-plane Re $y > 0$ and the half-strip $|\,\text{Im}\, y\,| < \frac{\pi}{2}$, Re $y \leq 0$. Moreover, by virtue of Eq. (4.B.4), the function defined in (4.B.7) reads

$$f(y) = \frac{z^2(4-z^2)}{4(1+z^2)^3}. \tag{4.B.9}$$

The non-trivial intermediate step is the double application of the rule for the differentiation of composite functions, which yields

$$\frac{d^2}{dy^2}(\dot{z}^{-1/2}) = \frac{(6z^2+1)}{4z^6}\left(1+\frac{1}{z^2}\right)^{-11/4}. \tag{4.B.10}$$

The result (4.B.9) implies that f is a holomorphic function of y in the domain D, since the points $z = \pm i$ do not belong to D. One also finds, from (4.B.8), that $z \sim y$ as $|\,y\,| \rightarrow \infty$ in the right-hand half-plane, whereas, if y lies in the strip $|\,\text{Im}\, y\,| < \frac{\pi}{2}$, Re $y \leq 0$, then

$$y \sim 1 + \log\frac{z}{2},$$

or, in terms of z,

$$z \sim 2e^{y-1},$$

as $|\,y\,| \rightarrow \infty$. Thus,

$$f(y) = \mathrm{O}(|\,y\,|^{-2}) \text{ as } |\,y\,| \rightarrow \infty \text{ in } D, \tag{4.B.11}$$

uniformly with respect to arg y.

Let now D' be the part of D bounded by the lines

$$\text{Im } y = \pm \left(\frac{\pi}{2} - \delta \right), \quad \text{Re } y \leq \delta,$$

$$\text{Re } y = \delta, \ | \text{ Im } y \ | \geq \frac{\pi}{2} - \delta,$$

with δ lying in the interval $]0, \frac{\pi}{2}]$. A theorem in the theory of asymptotic expansions (Olver 1954) guarantees now that, if one defines the sequence of coefficients $\{U_s(y)\}$ by means of the recurrence relations

$$U_0(y) = 1, \tag{4.B.12}$$

$$U_{s+1}(y) = -\frac{1}{2} U_s'(y) + \frac{1}{2} \int f(y) U_s(y) dy, \tag{4.B.13a}$$

two linearly independent solutions W_1 and W_2 of Eq. (4.B.6) exist such that, if y lies in D', their asymptotic expansions are given by

$$W_1(y) \sim e^{ny} \sum_{s=0}^{\infty} \frac{U_s(y)}{n^s}, \tag{4.B.14}$$

$$W_2(y) \sim e^{-ny} \sum_{s=0}^{\infty} (-1)^s \frac{U_s(y)}{n^s}, \tag{4.B.15}$$

as $n \to \infty$, *uniformly* with respect to y.

In the light of (4.B.1) and (4.B.5), one can express $I_n(nz)$ as the linear combination

$$I_n(nz) = \mu_n (1 + z^2)^{-1/4} W_1(y) + \rho_n (1 + z^2)^{-1/4} W_2(y), \tag{4.B.16}$$

where μ_n and ρ_n are independent of z. We may now take advantage of the uniform nature of the asymptotic expansions (4.B.14) and (4.B.15), and let $y \to -\infty$, while n is kept fixed. One has then to set $\rho_n = 0$ by comparison with the well-known behaviour of Bessel functions as the argument tends to zero. On combining (4.B.14) and (4.B.16) one then finds, for large n,

$$\mu_n \sim \frac{e^{-ny} (1 + z^2)^{1/4} I_n(nz)}{\sum_{s=0}^{\infty} \frac{U_s(y)}{n^s}}. \tag{4.B.17}$$

Further to this, it is known from the asymptotic expansion of Bessel functions at large argument and fixed order that

$$I_n(nz) \sim (2\pi nz)^{-1/2} e^{nz},$$

and hence one finds, as $z \to \infty$,

$$e^{-ny}(1+z^2)^{1/4} I_n(nz) \sim (2\pi n)^{-1/2}.$$

Last, on requiring that $U_{s+1}(\infty) = 0$, the arbitrary constant in Eq. (4.B.13a) is fixed:

$$U_{s+1}(y) = -\frac{1}{2} U_s'(y) + \frac{1}{2} \int_y^\infty f(t) U_s(t) dt, \tag{4.B.13b}$$

and one obtains the asymptotic behaviour (cf. (4.B.17))

$$\mu_n \sim (2\pi n)^{-1/2}. \tag{4.B.18}$$

By virtue of (4.B.14), (4.B.16) and (4.B.18) one finds that, as $n \to \infty$, the asymptotic expansion

$$I_n(nz) \sim \frac{e^{ny}}{\sqrt{2\pi n}(1+z^2)^{1/4}} \sum_{s=0}^{\infty} \frac{U_s(y)}{n^s} \tag{4.B.19}$$

holds *uniformly with respect to* z in the half-plane $|\arg z| \leq \frac{\pi}{2} - \varepsilon$, for ε in the open interval $]0, \frac{\pi}{2}[$.

For the modified Bessel function of the second kind, bearing in mind that, as $z \to \infty$,

$$K_n(nz) \sim \sqrt{\frac{\pi}{2nz}}\, e^{-nz},$$

one finds in an analogous way that, when $n \to \infty$,

$$K_n(nz) \sim \sqrt{\frac{\pi}{2n}} \frac{e^{-ny}}{(1+z^2)^{1/4}} \sum_{s=0}^{\infty} (-1)^s \frac{U_s(y)}{n^s} \tag{4.B.20}$$

is the uniform asymptotic expansion with respect to z in the half-plane $|\arg z| \leq \frac{\pi}{2} - \varepsilon$.

Of course, it would be desirable to have a recurrence formula which improves the ability of (4.B.13b) to generate the coefficients U_s in the uniform asymptotics. For this purpose, the form of (4.B.8) suggests we consider

$$u \equiv (1+z^2)^{-1/2}. \tag{4.B.21}$$

One then finds, by elementary differentiation,

$$\frac{du}{dy} = \frac{du}{dz}\frac{dz}{dy} = (u^4 - u^2), \tag{4.B.22}$$

$$\frac{dU_s}{dy} = \frac{dU_s}{du}\frac{du}{dy}, \tag{4.B.23}$$

$$f(y) = \frac{1}{4}u^2(1-u^2)(5u^2-1), \tag{4.B.24}$$

$$dy = \frac{du}{(u^4-u^2)}, \tag{4.B.25}$$

which imply, by virtue of (4.B.13*b*),

$$U_{s+1} = \frac{u^2}{2}(1-u^2)\frac{dU_s}{du} + \frac{1}{8}\int_0^u (1-5\beta^2)U_s(\beta)d\beta. \tag{4.B.26}$$

As far as the first derivatives of the Bessel functions $I_n(nz)$ and $K_n(nz)$ are concerned, their uniform asymptotic expansions, which are also very useful in the applications (Esposito *et al.* 1997) are obtained by differentiation of (4.B.19) and (4.B.20), and read (Olver 1954)

$$I'_n(nz) \sim \frac{(1+z^2)^{1/4}}{z}\frac{e^{ny}}{\sqrt{2\pi n}}\sum_{s=0}^{\infty}\frac{V_s}{n^s}, \tag{4.B.27}$$

$$K'_n(nz) \sim -\sqrt{\frac{\pi}{2n}}\frac{(1+z^2)^{1/4}}{z}e^{-ny}\sum_{s=0}^{\infty}(-1)^s\frac{V_s}{n^s}, \tag{4.B.28}$$

$$V_s = U_s - u(1-u^2)\left(\frac{1}{2}U_{s-1} + u\frac{dU_{s-1}}{du}\right). \tag{4.B.29}$$

5
Spectral geometry with operators of Laplace type

Spectral geometry with operators of Laplace type is introduced by discussing the inverse problem in the theory of vibrating membranes. This means that a spectrum of eigenvalues is given, and one would like to determine uniquely the shape of the vibrating object from the asymptotic expansion of the integrated heat kernel. It turns out that, in general, it is not possible to tell whether the membrane is convex, or smooth, or simply connected, but results of a limited nature can be obtained. These determine, for example, the volume and the surface area of the body. Starting from these examples, the very existence of the asymptotic expansion of the integrated heat hernel is discussed, relying on the seminal paper by Greiner (1971). A more careful analysis of the boundary-value problem is then performed, and the recent results on the asymptotics of the Laplacian on a manifold with boundary are presented in detail. For this purpose, one studies second-order elliptic operators with leading symbol given by the metric. The behaviour of the differential operator, boundary operator and heat-kernel coefficients under conformal rescalings of the background metric leads to a set of algebraic equations which, jointly with some results on product manifolds, determine a number of coefficients in the heat-kernel asymptotics. Such a property holds whenever one studies boundary conditions of Dirichlet or Robin type, or a mixture of the two. The heat-equation approach to index theorems, and the link between heat equation and ζ-function, are then described. The chapter ends with an introduction to the method used by McKean and Singer in their analysis of heat-kernel asymptotics.

5.1 On hearing the shape of a drum

A classical problem in spectral geometry, which is used here to introduce the topic, is the attempt to deduce the shape of a drum from the knowledge of its spectrum of eigenvalues, say λ_n. Some progress is possible on establishing the leading terms of the asymptotic expansion of the *trace function* (cf. section 5.2)

$$\Theta(t) = \sum_{n=1}^{\infty} e^{-\lambda_n t}, \tag{5.1.1}$$

for small positive values of t. In particular, if one studies a simply connected membrane Ω bounded by a smooth convex plane Γ, for which the displacement satisfies the wave equation (here, $\triangle \equiv -g^{ab}\nabla_a \nabla_b$, with g a Riemannian metric and ∇ the Levi–Civita connection compatible with g)

$$-\triangle \phi = \frac{\partial^2 \phi}{\partial t^2}, \tag{5.1.2}$$

and Dirichlet boundary conditions on Γ:

$$[\phi]_\Gamma = 0, \tag{5.1.3}$$

one finds

$$\Theta(t) \sim \frac{|\,\Omega\,|}{4\pi t} - \frac{L}{8\sqrt{\pi t}} + \frac{1}{6}. \tag{5.1.4}$$

In (5.1.4), $|\,\Omega\,|$ is the area of Ω, L is the length of Γ, and the constant $\frac{1}{6}$ results from integration of the curvature of the boundary. Moreover, if Ω has a finite number of smooth convex holes, such a term should be replaced by $\frac{1}{6}(1 - r)$, where r is the number of holes. The two basic problems in the heat-equation approach to drums and vibrating membranes are as follows (Kac 1966, Stewartson and Waechter 1971, Waechter 1972) (see also appendix 5.A).

(i) Given a set $\{\lambda_n\}$, can a corresponding shape be found? If so, is it unique?

(ii) Let $\{\lambda_n\}$ be a given spectrum of eigenvalues. Can the shape be uniquely determined from the asymptotic expansion of $\Theta(t)$ for small positive t?

In Stewartson and Waechter (1971), the authors derived the asymptotic expansion of $\Theta(t)$ for a membrane bounded by a smooth boundary. They found that, when the restriction of convexity is relaxed, and corners and cusps are permitted, the shape can be determined precisely if the membrane is circular; otherwise the asymptotic expansion of $\Theta(t)$ as $t \rightarrow 0^{+}$ determines the area, the length of the perimeter and the existence of outward pointing cusps. Thus, in general, *it is not possible* to tell whether the membrane is convex, or smooth, or simply connected. Whether it is possible to do so from the complete asymptotic expansion of $\Theta(t)$ remained, and remains, an open problem.

In the following sections we will give a rigorous description of heat-kernel methods, but here we can complete the outline of the physics-oriented approach, following again Stewartson and Waechter (1971). Indeed, if $G(r, r'; t)$ is the Green's function of the diffusion (or heat) equation

$$\left(\frac{\partial}{\partial t} + \triangle\right) \phi = 0, \tag{5.1.5}$$

subject to the Dirichlet condition

$$G(r, r'; t) = 0 \ \text{ if } \ r \in \Gamma, \tag{5.1.6}$$

and behaving as a Dirac delta: $\delta(r - r')$, as $t \rightarrow 0^{+}$, the trace function $\Theta(t)$ is defined by the equation (cf. section 5.2)

$$\Theta(t) \equiv \int \int_{\Omega} G(r, r; t) d\Omega. \tag{5.1.7}$$

Thus, bearing in mind that

$$G(r, r'; t) = \sum_{n=1}^{\infty} \phi_n(r) \phi_n(r') e^{-\lambda_n t}, \tag{5.1.8}$$

where λ_n are the eigenvalues of the operator $\triangle$, and ϕ_n are the normalized eigenfunctions, the result (5.1.1) follows on setting $r = r'$ and integrating over Ω. In many applications it is useful to consider the split

$$G(r, r'; t) = G_0(r, r'; t) + \chi(r, r'; t), \tag{5.1.9}$$

where (setting $R \equiv \mid r - r' \mid$)

$$G_0(r, r'; t) = \frac{1}{4\pi t} e^{-R^2/4t} \tag{5.1.10}$$

is the Green's function for the infinite plane, and $\chi(r, r'; t)$ is the compensating solution of the heat equation (5.1.5), which satisfies the appropriate boundary condition on Γ.

In Waechter (1972), the author extended the early investigation of the inverse eigenvalue problem for vibrating membranes to three or more dimensions. Thus, he considered the boundary-value problem

$$\Delta\phi = \lambda\,\phi \text{ in } \Omega, \tag{5.1.11}$$

$$\phi = 0 \text{ on } S, \tag{5.1.12}$$

where Ω is a closed convex region or body in the n-dimensional Euclidean space E^n, and S is the bounding surface of Ω. Once more, the problem was to determine the precise shape of Ω, on being given the spectrum of eigenvalues λ_n and the heat equation (5.1.5), whose Green's function satisfies the boundary condition

$$G(r, r'; t) = 0 \text{ if } r \in S. \tag{5.1.13}$$

The method was always to use the trace function

$$\Theta(t) \equiv \int\int\int_\Omega G(r, r; t)d\Omega, \tag{5.1.14}$$

to determine the leading terms of the asymptotic expansion of $\Theta(t)$ as $t \to 0^+$. Interestingly, even in the higher-dimensional problem, results of a limited nature can be obtained. For example, the first six terms of the expansion of $\Theta(t)$ for a sphere were determined, and for a smooth convex body Waechter found the first four terms:

$$\Theta(t) \sim \frac{V}{(4\pi t)^{3/2}} - \frac{S}{16\pi t} + \frac{\widetilde{M}}{6\pi\sqrt{4\pi t}} + \frac{1}{512\pi}\int\int_S (\kappa_1 - \kappa_2)^2 dS + O(\sqrt{t}). \tag{5.1.15}$$

With this notation, V and S are the volume and surface area of the body Ω, respectively, $\widetilde{M}$ is the surface integral of mean curvature:

$$\widetilde{M} \equiv \int\int \frac{1}{2}(\kappa_1 + \kappa_2)dS, \tag{5.1.16}$$

while κ_1 and κ_2 are the principal curvatures at the surface element dS of the body surface.

So far, nothing has been said about the actual *existence* of the asymptotic expansion of the trace function. The proof is a very important result due to Greiner (1971). Although we can only refer the reader to the original paper by Greiner for the detailed proof, one can describe, however, the mathematical framework, since this will help to achieve a smooth transition towards the following sections. We can thus say that Greiner studied a compact m-dimensional C^∞ Riemannian manifold, say M, with metric g and C^∞ boundary, say ∂M. The boundary-value problem consists of the pair (P, B), where P is an elliptic differential operator of order $2n$:

$$P : C^\infty(V, M) \rightarrow C^\infty(V, M), \tag{5.1.17}$$

with V a C^∞ (complex) vector bundle over M, and is such that $P + \frac{\partial}{\partial t}$ is parabolic. An example is the operator in round brackets in Eq. (5.1.5), with P being equal to $\triangle$. Moreover, the *boundary operator* is a map

$$B : C^\infty(V, M) \rightarrow C^\infty(\widetilde{V}, \partial M), \tag{5.1.18}$$

where $\widetilde{V}$ is a C^∞ (complex) vector bundle over ∂M. To obtain the Green's *kernel* $G(x, y; t)$ of e^{-tP}, Greiner constructed an operator C with kernel $C(x, y; t)$, compensating for the boundary, such that

$$G(x, y; t) = H(x, y; t) - C(x, y; t), \tag{5.1.19}$$

where $H(x, y; t)$ is the kernel of e^{-tP} for a manifold without boundary. Greiner's result states therefore that, for the *integrated (heat) kernel* (see section 5.2, and bear in mind that the integration is performed with respect to the Riemannian volume element $\sqrt{\det g}\, dx$)

$$G(t) \equiv \int_M \mathrm{Tr} G(x, x; t) = \mathrm{Tr}_{L^2}(e^{-tP}), \tag{5.1.20}$$

an asymptotic expansion exists, as $t \rightarrow 0^+$, in the form (cf. Greiner 1971)

$$G(t) \sim t^{-m/2n} \Big[G_0 + G_1 t^{1/2n} + \ldots + G_k t^{k/2n} + \ldots \Big], \tag{5.1.21}$$

where

$$G_0 = H_0, \tag{5.1.22}$$

$$G_k = H_k - C_k, \quad k = 1, 2, \ldots, \tag{5.1.23}$$

$$C_k = \int_{\partial M} C_k(x') dx'. \tag{5.1.24}$$

Note that the trace in the integral (5.1.20) is the fibre trace, and that in the mathematics literature one also says that $G(x, y; t)$ is the Green's kernel for the boundary-value problem $\left(P+\frac{\partial}{\partial t}, B\right)$. The following sections are devoted to a precise characterization of the G_k coefficients occurring in the asymptotic expansion (5.1.21).

5.2 The Laplacian on manifolds with boundary

Following Branson and Gilkey (1990), we are interested in a second-order differential operator, say P, with leading symbol given by the metric tensor on a compact m-dimensional Riemannian manifold M with boundary ∂M. Denoting by ∇ the connection on the vector bundle V, our assumption implies that P, called an operator of Laplace type, reads (cf. section 2.1)

$$P = -g^{ab}\nabla_a \nabla_b - E, \tag{5.2.1}$$

where E is an endomorphism of V. The heat equation for the operator P is (cf. (5.1.5))

$$\left(\frac{\partial}{\partial t} + P\right) U = 0. \tag{5.2.2}$$

By definition, the *heat kernel* (cf. section 5.1) is the solution, for $t > 0$, of the equation (Fulling 1989)

$$\left(\frac{\partial}{\partial t} + P\right) G(x, x'; t) = 0, \quad x \text{ and } x' \in M, \tag{5.2.3}$$

subject to the boundary condition

$$\Big[B\, G(x, x'; t)\Big]_{\partial M} = 0, \tag{5.2.4}$$

jointly with the (initial) condition (cf. (5.7.5))

$$\lim_{t \to 0^+} \int_M G(x, x'; t) \rho(x') dx' = \rho(x), \tag{5.2.5}$$

which is a rigorous mathematical expression for the Dirac delta behaviour as $t \to 0^+$ (cf. section 5.1). The heat kernel can be written as (cf. (5.1.8))

$$G(x, x'; t) = \sum_{(n)} \varphi_{(n)}(x) \varphi_{(n)}(x') e^{-\lambda_{(n)} t}, \tag{5.2.6}$$

where $\left\{\varphi_{(n)}(x)\right\}$ is a complete orthonormal set of eigenfunctions of P with eigenvalues $\lambda_{(n)}$. The index n is enclosed in round brackets, to emphasize that, in general, a finite collection of integer labels occurs therein.

Since, by construction, the heat kernel behaves as a distribution in the neighbourhood of the boundary, it is convenient to introduce a smooth function, say $f \in C^\infty(M)$, and consider a slight generalization of the trace function (or integrated heat kernel) of section 5.1, i.e. $\mathrm{Tr}_{L^2}\left(fe^{-tP}\right)$. It is precisely the consideration of f that makes it possible to deal properly with the distributional behaviour of the heat kernel near ∂M. A key idea is therefore to work with arbitrary f, and then set $f = 1$ only when all coefficients in the asymptotic expansion

$$\mathrm{Tr}_{L^2}\left(fe^{-tP}\right) \equiv \int_M \mathrm{Tr}\left[fG(x,x;t)\right]$$
$$\sim (4\pi t)^{-m/2} \sum_{n=0}^{\infty} t^{n/2} a_{n/2}(f,P,\mathcal{B}) \quad (5.2.7)$$

have been evaluated. The term $G(x,x;t)$ is called the *heat-kernel diagonal.* By virtue of Greiner's result, the heat-kernel coefficients $a_{n/2}(f,P,\mathcal{B})$, which are said to describe the *global asymptotics*, are obtained by integrating *local* formulae. More precisely, they admit a split into integrals over M (interior terms) and over ∂M (boundary terms). In such formulae, the integrands are linear combinations of all geometric invariants of the appropriate dimension (see below) which result from the Riemann curvature $R^a{}_{bcd}$ of the background, the extrinsic curvature of the boundary, the differential operator P (through the endomorphism E), and the boundary operator $\mathcal{B}$ (through the endomorphisms, or projection operators, or more general matrices occurring in it). With our notation, the indices $a, b, \ldots$ range from 1 through m and index a local orthonormal frame for the tangent bundle of M, TM, while the indices $i, j, \ldots$ range from 1 through $m-1$ and index the orthonormal frame for the tangent bundle of the boundary, $T(\partial M)$. The boundary is defined by the equations

$$\partial M: \qquad y^a = y^a(x), \quad (5.2.8)$$

in terms of the functions $y^a(x)$, x^i being the coordinates on ∂M, and the y^a those on M. Thus, the *intrinsic* metric, γ_{ij}, on the boundary hypersurface ∂M, is given in terms of the metric g_{ab} on M by (Eisenhart 1926, Dowker and Schofield 1990)

$$\gamma_{ij} = g_{ab}\, y^a{}_{,i}\, y^b{}_{,j}. \quad (5.2.9)$$

On inverting this equation one finds (here, $n^a = N^a$ is the inward-pointing normal)

$$g^{ab} = q^{ab} + n^a n^b, \tag{5.2.10}$$

where

$$q^{ab} = y^a_{\ ,i}\, y^b_{\ ,j}\, \gamma^{ij}. \tag{5.2.11}$$

The tensor q^{ab} is equivalent to γ^{ij} and may be viewed as the *induced metric* on ∂M, in its contravariant form. The tensor $q^a_{\ b}$ is a projection operator, in that

$$q^a_{\ b}\, q^b_{\ c} = q^a_{\ c}, \tag{5.2.12}$$

$$q^a_{\ b}\, n^b = 0. \tag{5.2.13}$$

The extrinsic-curvature tensor K_{ab} (or second fundamental form of ∂M) is here defined by the projection of the covariant derivative of an *extension* of the inward, normal vector field n:

$$K_{ab} \equiv n_{c;d}\, q^c_{\ a}\, q^d_{\ b}, \tag{5.2.14}$$

and is symmetric if the metric-compatible connection on M is torsion-free. Only its spatial components, K_{ij}, are non-vanishing.

The semicolon ; denotes multiple covariant differentiation with respect to the Levi–Civita connection ∇_M of M, while the stroke | denotes multiple covariant differentiation tangentially with respect to the Levi–Civita connection $\nabla_{\partial M}$ of the boundary. When sections of bundles built from V are involved, the semicolon means

$$\nabla_M \otimes \mathbb{1} + \mathbb{1} \otimes \nabla,$$

and the stroke means

$$\nabla_{\partial M} \otimes \mathbb{1} + \mathbb{1} \otimes \nabla.$$

The curvature of the connection ∇ on V is denoted by Ω.

When Dirichlet or Robin boundary conditions are imposed on sections of V:

$$[\phi]_{\partial M} = 0, \tag{5.2.15}$$

or

$$\Big[(n^a \nabla_a + S)\phi\Big]_{\partial M} = 0, \tag{5.2.16}$$

the asymptotics in (5.2.7) is expressed through some *universal constants*

$$\{\alpha_i, b_i, c_i, d_i, e_i\}$$

such that (here the Riemann tensor coincides with the definition (1.2.1), $R_{ab} \equiv R^c{}_{abc}$ is the Ricci tensor, $R \equiv R^a{}_a$, and $\Box \equiv \nabla^a \nabla_a = g^{ab} \nabla_a \nabla_b$)

$$a_0(f, P, \mathcal{B}) = \int_M \mathrm{Tr}(f), \tag{5.2.17}$$

$$a_{1/2}(f, P, \mathcal{B}) = \delta(4\pi)^{1/2} \int_{\partial M} \mathrm{Tr}(f), \tag{5.2.18}$$

$$a_1(f, P, \mathcal{B}) = \frac{1}{6} \int_M \mathrm{Tr}\Big[\alpha_1 f E + \alpha_2 f R\Big] + \frac{1}{6} \int_{\partial M} \mathrm{Tr}\Big[b_0 f (\mathrm{tr}K) + b_1 f_{;N} + b_2 f S\Big], \tag{5.2.19}$$

$$a_{3/2}(f, P, \mathcal{B}) = \frac{\delta}{96}(4\pi)^{1/2} \int_{\partial M} \mathrm{Tr}\Big[f\Big(c_0 E + c_1 R + c_2 R^i{}_{NiN} + c_3 (\mathrm{tr}K)^2 + c_4 K_{ij} K^{ij} + c_7 S (\mathrm{tr}K) + c_8 S^2\Big) + f_{;N}\Big(c_5 (\mathrm{tr}K) + c_9 S\Big) + c_6 f_{;NN}\Big], \tag{5.2.20}$$

$$\begin{aligned} a_2(f, P, \mathcal{B}) = {} & \frac{1}{360} \int_M \mathrm{Tr}\Big[f\Big(\alpha_3 \Box E + \alpha_4 R E + \alpha_5 E^2 + \alpha_6 \Box R + \alpha_7 R^2 + \alpha_8 R_{ab} R^{ab} \\ & + \alpha_9 R_{abcd} R^{abcd} + \alpha_{10} \Omega_{ab} \Omega^{ab}\Big)\Big] \\ & + \frac{1}{360} \int_{\partial M} \mathrm{Tr}\Big[f\Big(d_1 E_{;N} + d_2 R_{;N} + d_3 (\mathrm{tr}K)_{|i}{}^{|i} + d_4 K_{ij}{}^{|ij} \\ & + d_5 E (\mathrm{tr}K) + d_6 R (\mathrm{tr}K) + d_7 R^i{}_{NiN} (\mathrm{tr}K) + d_8 R_{iNjN} K^{ij} \\ & + d_9 R^l{}_{ilj} K^{ij} + d_{10} (\mathrm{tr}K)^3 + d_{11} K_{ij} K^{ij} (\mathrm{tr}K) \\ & + d_{12} K_i{}^j K_j{}^l K_l{}^i + d_{13} \Omega_{iN;}{}^i + d_{14} S E \\ & + d_{15} S R + d_{16} S R^i{}_{NiN} + d_{17} S (\mathrm{tr}K)^2 + d_{18} S K_{ij} K^{ij} \\ & + d_{19} S^2 (\mathrm{tr}K) + d_{20} S^3 + d_{21} S_{|i}{}^{|i}\Big) + f_{;N}\Big(e_1 E + e_2 R \\ & + e_3 R^i{}_{NiN} + e_4 (\mathrm{tr}K)^2 + e_5 K_{ij} K^{ij} + e_8 S (\mathrm{tr}K) + e_9 S^2\Big) \\ & + f_{;NN}\Big(e_6 (\mathrm{tr}K) + e_{10} S\Big) + e_7 f_{;a}{}^a{}_N\Big]. \end{aligned} \tag{5.2.21}$$

These formulae may seem to be very complicated, but there is indeed a systematic way to write them down and then compute the universal constants. To begin, note that, if k is odd, $a_{k/2}(f, P, \mathcal{B})$ receives contributions from boundary terms only, whereas both interior terms and boundary terms contribute to $a_{k/2}(f, P, \mathcal{B})$, if k is even and positive. In the a_1 coefficient, the integrand in the interior term must be linear in the curvature, and hence it can only be a linear combination of the trace of the Ricci tensor, and of the endomorphism E in the differential operator. In the a_2 coefficient, the integrand in the interior term must be quadratic in the curvature, and hence one needs a linear combination of the eight geometric invariants (cf. Fock 1937, Schwinger 1951, Hadamard 1952, DeWitt 1965)

$$\Box E \ , \ RE \ , \ E^2 \ , \ \Box R \ , \ R^2 \ , \ R_{ab}R^{ab} \ , \ R_{abcd}R^{abcd} \ , \ \Omega_{ab}\Omega^{ab}.$$

In the a_1 coefficient, the integrand in the boundary term is a local expression given by a linear combination of all invariants linear in the extrinsic curvature: $\mathrm{tr}K, S$ and $f_{;N}$. In the $a_{3/2}$ coefficient, the integrand in the boundary term must be quadratic in the extrinsic curvature. Thus, bearing in mind the Gauss–Codazzi equations, one finds the general result (5.2.20). Last, in the a_2 coefficient, the integrand in the boundary term must be cubic in the extrinsic curvature. This leads to the boundary integral in (5.2.21), bearing in mind that $f_{;N}$ is linear in K_{ij}, while $f_{;NN}$ is quadratic in K_{ij}. Note that the interior invariants are built universally and polynomially from the metric tensor, its inverse, and the covariant derivatives of $R^a_{\ bcd}, \Omega_{ab}$ and E. By virtue of Weyl's work on the invariants of the orthogonal group (Weyl 1946, Branson and Gilkey 1990, Gilkey 1995), these polynomials can be formed using only tensor products and contraction of tensor arguments. Here, the structure group is $O(m)$. However, when a boundary occurs, the boundary structure group is $O(m-1)$. Weyl's theorem is used again to construct invariants as in the previous equations (Branson and Gilkey 1990).

5.3 Functorial method

Let T be a map which carries finite-dimensional vector spaces into finite-dimensional vector spaces. Thus, to every vector space V one has an associated vector space $T(V)$. The map T is said to be a *continuous functor* if, for all V and W, the map

$$T : Hom(V,W) \longrightarrow Hom(T(V),T(W))$$

is continuous (Atiyah 1967).

In the theory of heat kernels, an application of the functorial method is the analysis of heat-equation asymptotics with respect to conformal variations. Indeed, the behaviour of classical and quantum field theories under conformal rescalings of the metric

$$\widehat{g}_{ab} = \Omega^2 \, g_{ab}, \tag{5.3.1}$$

with Ω a smooth function, is at the heart of many deep properties: light-cone structure, conformal curvature (i.e. the Weyl tensor), conformal-infinity techniques, massless free-field equations, twistor equation, twistor spaces, Hodge-star operator in four dimensions, conformal anomalies (Penrose and Rindler 1986, Ward and Wells 1990, Esposito 1994a, 1995, Esposito *et al.* 1997). In the functorial method, one chooses Ω in the form

$$\Omega = e^{\varepsilon f}, \tag{5.3.2}$$

where ε is a real-valued parameter, and $f \in C^\infty(M)$ is the smooth function considered in section 5.2. One then deals with a one-parameter family of differential operators

$$P(\varepsilon) = e^{-2\varepsilon f} \, P(0), \tag{5.3.3}$$

boundary operators

$$\mathcal{B}(\varepsilon) = e^{-\varepsilon f} \, \mathcal{B}(0), \tag{5.3.4}$$

connections ∇^ε on V, endomorphisms $E(\varepsilon)$ of V, and metrics

$$g_{ab}(\varepsilon) = e^{2\varepsilon f} \, g_{ab}(0). \tag{5.3.5}$$

For example, the form (5.3.2) of the conformal factor should be inserted into the general formulae which describe the transformation of Christoffel symbols under conformal rescalings:

$$\widehat{\Gamma}^a_{\ bc} = \Gamma^a_{\ bc} + \Omega^{-1}\Big(\delta^a_{\ b} \, \Omega_{,c} + \delta^a_{\ c} \, \Omega_{,b} - g_{bc} g^{ad} \Omega_{,d}\Big). \tag{5.3.6}$$

This makes it possible to obtain the conformal-variation formulae for the Riemann tensor $R^a_{\ bcd}$ and for all tensors involving the effect of Christoffel symbols. For the extrinsic-curvature tensor defined in Eq. (5.2.14) one finds

$$\widehat{K}_{ab} = \Omega K_{ab} - n_a \nabla_b \Omega + g_{ab} \nabla_{(n)} \Omega, \tag{5.3.7}$$

which implies

$$K_{ij}(\varepsilon) = e^{\varepsilon f}\, K_{ij}(0) - \varepsilon\, g_{ij}\, e^{\varepsilon f}\, f_{;N}. \tag{5.3.8}$$

The application of these methods to heat-kernel asymptotics relies on the work by Branson *et al.* (1990) and Branson and Gilkey (1990). Within this framework, a crucial role is played by the following 'functorial' formulae (F being another smooth function):

$$\left[\frac{d}{d\varepsilon} a_{n/2}\Big(1, e^{-2\varepsilon f} P(0)\Big)\right]_{\varepsilon=0} = (m-n) a_{n/2}(f, P(0)), \tag{5.3.9}$$

$$\left[\frac{d}{d\varepsilon} a_{n/2}\Big(1, P(0) - \varepsilon F\Big)\right]_{\varepsilon=0} = a_{n/2-1}(F, P(0)), \tag{5.3.10}$$

$$\left[\frac{d}{d\varepsilon} a_{n/2}\Big(e^{-2\varepsilon f} F, e^{-2\varepsilon f} P(0)\Big)\right]_{\varepsilon=0} = 0. \tag{5.3.11}$$

Equation (5.3.11) is obtained when $m = n + 2$. These properties can be proved by (formal) differentiation, as follows.

If the conformal variation of an operator of Laplace type reads

$$P(\varepsilon) = e^{-2\varepsilon f} P(0) - \varepsilon F, \tag{5.3.12}$$

one finds

$$\left[\frac{d}{d\varepsilon} \mathrm{Tr}_{L^2}\Big(e^{-tP(\varepsilon)}\Big)\right]_{\varepsilon=0} = \mathrm{Tr}_{L^2}\left[\Big(2tfP(0) + tF\Big) e^{-tP(0)}\right]$$

$$= -2t \frac{\partial}{\partial t} \mathrm{Tr}_{L^2}\Big(f\, e^{-tP(0)}\Big) + t\,\mathrm{Tr}_{L^2}\Big(F\, e^{-tP(0)}\Big). \tag{5.3.13}$$

Moreover, by virtue of the asymptotic expansion (5.2.7), one has (the numerical factors $(4\pi)^{-m/2}$ are omitted for simplicity, since they do not affect the form of Eqs. (5.3.9)–(5.3.11); following Branson and Gilkey (1990) one can, instead, absorb such factors into the definition of the coefficients $a_{n/2}(f, P)$)

$$\frac{\partial}{\partial t} \mathrm{Tr}_{L^2}\Big(f\, e^{-tP(0)}\Big) \sim -\frac{1}{2} \sum_{n=0}^{\infty} (m-n) t^{\frac{n}{2} - \frac{m}{2} - 1}\, a_{n/2}(f, P(0)). \tag{5.3.14}$$

Thus, if F vanishes, Eqs. (5.3.13) and (5.3.14) lead to the result (5.3.9). By contrast, if f is set to zero, one has $P(\varepsilon) = P(0) - \varepsilon F$, which implies

$$\left[\frac{d}{d\varepsilon} \mathrm{Tr}_{L^2}\Big(e^{-tP(\varepsilon)}\Big)\right]_{\varepsilon=0} \sim t^{-m/2} \sum_{n=0}^{\infty} t^{n/2+1}\, a_{n/2}(F, P(0))$$

$$= t^{-m/2} \sum_{l=2}^{\infty} t^{l/2} \, a_{l/2-1}(F, P(0)), \tag{5.3.15}$$

which leads in turn to Eq. (5.3.10). Last, to obtain the result (5.3.11), one considers the two-parameter conformal variation

$$P(\varepsilon, \gamma) = e^{-2\varepsilon f} P(0) - \gamma e^{-2\varepsilon f} F. \tag{5.3.16}$$

Now in Eq. (5.3.9) we first replace n by $n+2$, and then set $m = n+2$. One then has, from Eq. (5.3.16):

$$\frac{\partial}{\partial \varepsilon} a_{n/2+1}(1, P(\varepsilon, \gamma)) = 0. \tag{5.3.17}$$

Equation (5.3.17) can be differentiated with respect to γ, i.e. (see (5.3.10))

$$0 = \frac{\partial^2}{\partial \gamma \partial \varepsilon} a_{n/2+1}(1, P(\varepsilon, \gamma)) = \frac{\partial}{\partial \varepsilon} \frac{\partial}{\partial \gamma} a_{n/2+1}\Big(1, e^{-2\varepsilon f}(P(0) - \gamma F)\Big)$$
$$= \frac{\partial}{\partial \varepsilon} a_{n/2}\Big(e^{-2\varepsilon f} F, e^{-2\varepsilon f} P(0)\Big), \tag{5.3.18}$$

and hence Eq. (5.3.11) is proved.

To deal with Robin boundary conditions (called Neumann in Branson and Gilkey (1990)) one needs another lemma, which is proved following again Branson and Gilkey. What we obtain is indeed a particular case of a more general property, which is proved in section 6 of Avramidi and Esposito (1998a). Our starting point is M, a compact, connected one-dimensional Riemannian manifold with boundary. In other words, one deals with the circle or with a closed interval. If

$$b : C^\infty(M) \longrightarrow \Re$$

is a smooth, real-valued function, one can form the first-order operator

$$A \equiv \frac{d}{dx} - b, \tag{5.3.19}$$

and its (formal) adjoint

$$A^\dagger \equiv -\frac{d}{dx} - b. \tag{5.3.20}$$

From these operators, one can form the second-order operators (cf. sections 5.5 and 5.6)

$$D_1 \equiv A^\dagger A = -\left[\frac{d^2}{dx^2} - b_x - b^2\right], \tag{5.3.21}$$

$$D_2 \equiv AA^\dagger = -\left[\frac{d^2}{dx^2} + b_x - b^2\right], \tag{5.3.22}$$

where $b_x \equiv \frac{db}{dx}$. For D_1, Dirichlet boundary conditions are taken, while Robin boundary conditions are assumed for D_2. Defining

$$f_x \equiv \frac{df}{dx}, \ f_{xx} \equiv \frac{d^2 f}{dx^2},$$

one then finds the result (cf. Branson and Gilkey 1990):

$$(n-1)\Big[a_{n/2}(f, D_1) - a_{n/2}(f, D_2)\Big] = a_{n/2-1}\Big(f_{xx} + 2b f_x, D_1\Big). \tag{5.3.23}$$

As a first step in the proof of (5.3.23), one takes a spectral resolution for D_1, say $\{\theta_\nu, \lambda_\nu\}$, where θ_ν is the eigenfunction of D_1 belonging to the eigenvalue λ_ν:

$$D_1 \, \theta_\nu = \lambda_\nu \, \theta_\nu. \tag{5.3.24}$$

Thus, differentiation with respect to t of the heat-kernel diagonal:

$$K(D_1, x, x; t) = \sum_\nu e^{-t\lambda_\nu} \, \theta_\nu^2(x), \tag{5.3.25}$$

yields

$$\frac{\partial}{\partial t} K(D_1, x, x; t) = -\sum_\nu \lambda_\nu e^{-t\lambda_\nu} \, \theta_\nu^2(x) = -\sum_\nu e^{-t\lambda_\nu} (D_1 \, \theta_\nu)\theta_\nu. \tag{5.3.26}$$

Moreover, for any $\lambda_\nu \neq 0$, the set

$$\left\{\frac{A\theta_\nu}{\sqrt{\lambda_\nu}}, \lambda_\nu\right\}$$

provides a spectral resolution of D_2 on $\mathrm{Ker}(D_2)^\perp$, and one finds

$$\begin{aligned}\frac{\partial}{\partial t} K(D_2, x, x; t) &= -\sum_{\lambda_\nu \neq 0} \lambda_\nu e^{-t\lambda_\nu} \, \theta_\nu^2(x) \\ &= -\sum_{\lambda_\nu \neq 0} e^{-t\lambda_\nu} \Big(\sqrt{\lambda_\nu} \, \theta_\nu\Big)\Big(\sqrt{\lambda_\nu} \, \theta_\nu\Big) \\ &= -\sum_{\lambda_\nu \neq 0} e^{-t\lambda_\nu} (A\theta_\nu)(A\theta_\nu).\end{aligned} \tag{5.3.27}$$

Bearing in mind that $A\theta_\nu = 0$ if $\lambda_\nu = 0$, summation may be performed over all values of ν, to find

$$2\frac{\partial}{\partial t}\Big[K(D_1, x, x; t) - K(D_2, x, x; t)\Big]$$

$$= 2\sum_\nu e^{-t\lambda_\nu}\Big[\Big(\theta_\nu''\theta_\nu - b'\theta_\nu^2 - b^2\theta_\nu^2\Big) + (\theta_\nu' - b\theta_\nu)(\theta_\nu' - b\theta_\nu)\Big]$$

$$= 2\sum_\nu e^{-t\lambda_\nu}\Big[\theta_\nu''\theta_\nu - b'\theta_\nu^2 + (\theta_\nu')^2 - 2b\theta_\nu'\theta_\nu\Big]. \qquad (5.3.28)$$

On the other hand, differentiation with respect to x yields

$$\left(\frac{\partial}{\partial x} - 2b\right) K(D_1, x, x; t) = \sum_\nu e^{-t\lambda_\nu}\Big(2\theta_\nu\theta_\nu' - 2b\theta_\nu^2\Big), \qquad (5.3.29)$$

which implies

$$f\frac{\partial}{\partial x}\left(\frac{\partial}{\partial x} - 2b\right) K(D_1, x, x; t)$$

$$= 2f\frac{\partial}{\partial t}\Big[K(D_1, x, x; t) - K(D_2, x, x; t)\Big]. \qquad (5.3.30)$$

We now integrate this formula over M and use the boundary conditions described before, jointly with integration by parts. All boundary terms are found to vanish, so that

$$\int_M f\frac{\partial}{\partial x}\left(\frac{\partial}{\partial x} - 2b\right) K(D_1, x, x; t)dx$$

$$= \int_M \left(\frac{\partial^2 f}{\partial x^2} + 2b\frac{\partial f}{\partial x}\right) K(D_1, x, x; t)dx$$

$$= \int_M 2f\frac{\partial}{\partial t}\Big[K(D_1, x, x; t) - K(D_2, x, x; t)\Big]. \qquad (5.3.31)$$

Bearing in mind the standard notation for heat-kernel traces, Eq. (5.3.31) may be re-expressed as

$$2\frac{\partial}{\partial t}\Big[\mathrm{Tr}_{L^2}\Big(fe^{-tD_1}\Big) - \mathrm{Tr}_{L^2}\Big(fe^{-tD_2}\Big)\Big]$$

$$= \mathrm{Tr}_{L^2}\Big[\Big(f_{xx} + 2bf_x\Big)e^{-tD_1}\Big], \qquad (5.3.32)$$

which leads to Eq. (5.3.23) by virtue of the asymptotic expansion (5.2.7).

The algorithm resulting from Eq. (5.3.10) is sufficient to determine almost all interior terms in heat-kernel asymptotics. To appreciate this, notice that one is dealing with conformal variations which only affect the endomorphism of the operator P in (5.2.1). For example, on setting $n = 2$ in Eq. (5.3.10), one ends up by studying (the tilde symbol is now used for interior terms)

$$\tilde{a}_1(1, P) \equiv \frac{1}{6}\int_M \mathrm{Tr}\Big[\alpha_1 E + \alpha_2 R\Big] = \tilde{a}_1(E, R), \qquad (5.3.33)$$

which implies

$$\tilde{a}_1(1, P(0) - \varepsilon F) = \tilde{a}_1(E, R) - \tilde{a}_1(E - \varepsilon F, R)$$
$$= \frac{1}{6} \int_M \mathrm{Tr}\Big[\alpha_1(E - (E - \varepsilon F)) + \alpha_2(R - R)\Big]$$
$$= \frac{1}{6} \int_M \mathrm{Tr}(\alpha_1 \varepsilon F), \qquad (5.3.34)$$

and hence

$$\left[\frac{d}{d\varepsilon}\tilde{a}_1(1, P(0) - \varepsilon F)\right]_{\varepsilon=0} = \frac{1}{6} \int_M \mathrm{Tr}(\alpha_1 F)$$
$$= \tilde{a}_0(F, P(0)) = \int_M \mathrm{Tr}(F). \qquad (5.3.35)$$

By comparison, Eq. (5.3.35) shows that

$$\alpha_1 = 6. \qquad (5.3.36)$$

An analogous procedure leads to (see (5.2.21))

$$\tilde{a}_2(1, P(0) - \varepsilon F) = \tilde{a}_2\Big(E, R, \mathrm{Ric}, \mathrm{Riem}, \Omega\Big)$$
$$- \tilde{a}_2\Big(E - \varepsilon F, R, \mathrm{Ric}, \mathrm{Riem}, \Omega\Big)$$
$$= \frac{1}{360} \int_M \mathrm{Tr}\Big[\alpha_4 R \varepsilon F + \alpha_5(E^2 - (E - \varepsilon F)^2)\Big]$$
$$= \frac{1}{360} \int_M \mathrm{Tr}\Big[\alpha_4 R \varepsilon F + \alpha_5\Big(-\varepsilon^2 F^2 + 2\varepsilon E F\Big)\Big]. \qquad (5.3.37)$$

Now one can apply Eq. (5.3.10) when $n = 4$, to find

$$\left[\frac{d}{d\varepsilon}\tilde{a}_2(1, P(0) - \varepsilon F)\right]_{\varepsilon=0} = \frac{1}{360} \int_M \mathrm{Tr}\Big[\alpha_4 F R + 2\alpha_5 F E\Big]$$
$$= \tilde{a}_1(F, P(0)) = \frac{1}{6} \int_M \mathrm{Tr}\Big[\alpha_1 F E + \alpha_2 F R\Big]. \qquad (5.3.38)$$

Equating the coefficients of the invariants occurring in the equation (5.3.38) one finds

$$\alpha_2 = \frac{1}{60}\alpha_4, \qquad (5.3.39)$$

$$\alpha_5 = 30\alpha_1 = 180. \qquad (5.3.40)$$

Further, the consideration of Eq. (5.3.11) when $n = 2$ yields

$$\left[\frac{d}{d\varepsilon}\tilde{a}_1\Big(e^{-2\varepsilon f} F, e^{-2\varepsilon f} P(0)\Big)\right]_{\varepsilon=0} = \frac{1}{6} \int_M \left\{\left[\frac{d}{d\varepsilon}\mathrm{Tr}(\alpha_1 F E)\right]_{\varepsilon=0}\right.$$

$$+ 2f\mathrm{Tr}(\alpha_1 FE) + \left[\frac{d}{d\varepsilon}\mathrm{Tr}(\alpha_2 FR)\right]_{\varepsilon=0} + 2f\mathrm{Tr}(\alpha_2 FR)\Big\}. \quad (5.3.41)$$

At this stage, we need the conformal-variation formulae

$$\left[\frac{d}{d\varepsilon}E(\varepsilon)\right]_{\varepsilon=0} = -2fE + \frac{1}{2}(m-2)\Box f, \quad (5.3.42)$$

$$\left[\frac{d}{d\varepsilon}R(\varepsilon)\right]_{\varepsilon=0} = -2fR - 2(m-1)\Box f. \quad (5.3.43)$$

Since we are studying the case $m = n + 2 = 4$, we find

$$\left[\frac{d}{d\varepsilon}\tilde{a}_1\left(e^{-2\varepsilon f}F, e^{-2\varepsilon f}P(0)\right)\right]_{\varepsilon=0} = \frac{1}{6}\int_M \mathrm{Tr}\Big[(\alpha_1 - 6\alpha_2)F\Box f\Big] = 0, \quad (5.3.44)$$

which implies

$$\alpha_2 = \frac{1}{6}\alpha_1 = 1, \quad (5.3.45)$$

$$\alpha_4 = 60\alpha_2 = 60. \quad (5.3.46)$$

After considering the Laplacian acting on functions for a product manifold $M = M_1 \times M_2$ (this is another application of functorial methods), the complete set of coefficients for interior terms can be determined (Branson and Gilkey 1990):

$$\alpha_3 = 60\ ,\ \alpha_6 = 12\ ,\ \alpha_7 = 5\ ,\ \alpha_8 = -2\ ,\ \alpha_9 = 2\ ,\ \alpha_{10} = 30. \quad (5.3.47)$$

As far as interior terms are concerned, one has to use Eqs. (5.3.9) and (5.3.11), jointly with two conformal-variation formulae which provide divergence terms that play an important role (Branson and Gilkey 1990):

$$\left[\frac{d}{d\varepsilon}a_1\left(F, e^{-2\varepsilon f}P(0)\right)\right]_{\varepsilon=0} - (m-2)a_1(fF, P(0))$$
$$= \frac{1}{6}(m-4)\int_M \mathrm{Tr}\Big(F\Box f\Big), \quad (5.3.48)$$

$$\left[\frac{d}{d\varepsilon}a_2\left(F, e^{-2\varepsilon f}P(0)\right)\right]_{\varepsilon=0} - (m-4)a_2(fF, P(0))$$
$$= \frac{1}{360}(m-6)\int_M F\mathrm{Tr}\Big[6f^{;b}{}_{;b}{}^{a} + 10f^{;a}R$$
$$+ 60f^{;a}E + 4f_{;b}R^{ab}\Big]_{;a}. \quad (5.3.49)$$

Equations (5.3.48) and (5.3.49) are proved for manifolds without boundary in Lemma 4.2 of Branson and Gilkey (1990). They imply that, for manifolds with boundary, the right-hand side of Eq. (5.3.48), evaluated for $F = 1$, should be added to the left-hand side of Eq. (5.3.9) when $n = 2$. Similarly, the right-hand side of Eq. (5.3.49), evaluated for $F = 1$, should be added to the left-hand side of Eq. (5.3.9) when $n = 4$. Other useful formulae involving boundary effects are (Branson and Gilkey 1990)

$$\int_M \left(f_{;b} R^{ab}\right)_{;a} = \int_{\partial M} \left[f_{;j}\left(K^{ij}{}_{|i} - (\mathrm{tr}K)^{|j}\right) + f_{;N} R^i{}_{NiN}\right], \quad (5.3.50)$$

$$f_{;i}{}^{;i} = f_{|i}{}^{|i} - (\mathrm{tr}K) f_{;N}, \quad (5.3.51)$$

$$\int_M \Box f = - \int_{\partial M} f_{;N}, \quad (5.3.52)$$

$$\int_{\partial M} \Box f = \int_{\partial M} \left[f_{;NN} - (\mathrm{tr}K) f_{;N}\right]. \quad (5.3.53)$$

On taking into account Eqs. (5.3.48)–(5.3.53), the application of Eq. (5.3.9) when $n = 2, 3, 4$ leads to 18 equations which are obtained by setting to zero the coefficients multiplying

$$f_{;N} \ \ (\text{when } n = 2),$$

$$f_{;NN} \ , \ f_{;N}(\mathrm{tr}K) \ , \ f_{;N} S \ \ (\text{when } n = 3),$$

and

$$f_{;a}{}^{;a}{}_N \ , \ f_{;N} E \ , \ f_{;N} R \ , \ f_{;N} R^i{}_{NiN},$$

$$f_{;NN}(\mathrm{tr}K) \ , \ f_{;N}(\mathrm{tr}K)^2 \ , \ f_{;N} K_{ij} K^{ij} \ , \ f_{|i}(\mathrm{tr}K)^{|i},$$

$$f_{|i} K^{ij}{}_{|j} \ , \ f_{|i} \Omega^i{}_N \ , \ f_{;N} S(\mathrm{tr}K) \ , \ f_{;N} S^2,$$

$$f_{;NN} S \ , \ f_{|i}{}^{|i} S \ \ (\text{when } n = 4).$$

The integrals of these 18 terms have a deep geometric nature in that they form a basis for the integral invariants. The resulting 18 equations are

$$-b_0(m-1) - b_1(m-2) + \frac{1}{2} b_2(m-2) - (m-4) = 0, \quad (5.3.54)$$

$$\frac{1}{2} c_0(m-2) - 2c_1(m-1) + c_2(m-1) - c_6(m-3) = 0, \quad (5.3.55)$$

$$-\frac{1}{2}c_0(m-2)+2c_1(m-1)-c_2-2c_3(m-1)$$
$$-2c_4-c_5(m-3)+\frac{1}{2}c_7(m-2)=0, \tag{5.3.56}$$

$$-c_7(m-1)+c_8(m-2)-c_9(m-3)=0, \tag{5.3.57}$$

$$-6(m-6)+\frac{1}{2}d_1(m-2)-2d_2(m-1)-e_7(m-4)=0, \tag{5.3.58}$$

$$-60(m-6)-2d_1-d_5(m-1)+\frac{1}{2}d_{14}(m-2)-e_1(m-4)=0, \tag{5.3.59}$$

$$-10(m-6)-2d_2-d_6(m-1)$$
$$+d_9+\frac{1}{2}d_{15}(m-2)-e_2(m-4)=0, \tag{5.3.60}$$

$$-d_7(m-1)-d_8+2d_9-e_3(m-4)$$
$$+\frac{1}{2}d_{16}(m-2)+4(m-6)=0, \tag{5.3.61}$$

$$\frac{1}{2}d_5(m-2)-2d_6(m-1)+d_7(m-1)+d_8-e_6(m-4)=0, \tag{5.3.62}$$

$$-\frac{1}{2}d_5(m-2)+2d_6(m-1)-d_7-d_9-3d_{10}(m-1)$$
$$-2d_{11}+\frac{1}{2}d_{17}(m-2)-e_4(m-4)=0, \tag{5.3.63}$$

$$-d_8-d_9(m-3)-d_{11}(m-1)-3d_{12}$$
$$+\frac{1}{2}d_{18}(m-2)-e_5(m-4)=0, \tag{5.3.64}$$

$$d_3(m-4)-\frac{1}{2}d_5(m-2)+2d_6(m-1)-d_7-d_9-4(m-6)=0, \tag{5.3.65}$$

$$d_4(m-4)-d_8-d_9(m-3)+4(m-6)=0, \tag{5.3.66}$$

$$(m-4)d_{13}=0, \tag{5.3.67}$$

$$-\frac{1}{2}d_{14}(m-2)+2d_{15}(m-1)-d_{16}-2d_{17}(m-1)$$
$$-2d_{18}+d_{19}(m-2)-e_8(m-4)=0, \tag{5.3.68}$$

$$-d_{19}(m-1)+\frac{3}{2}d_{20}(m-2)-e_9(m-4)=0, \tag{5.3.69}$$

$$\frac{1}{2}d_{14}(m-2)-2d_{15}(m-1)+d_{16}(m-1)-e_{10}(m-4)=0, \quad (5.3.70)$$

$$\frac{1}{2}d_{14}(m-2)-2d_{15}(m-1)+d_{16}-d_{21}(m-4)=0. \quad (5.3.71)$$

This set of algebraic equations among universal constants holds independently of the choice of Dirichlet or Robin boundary conditions. Another set of equations which hold for either Dirichlet or Robin boundary conditions is obtained by applying Eq. (5.3.11) when $n=3,4$. One then obtains five equations which result from setting to zero the coefficients multiplying

$$F_{;N}f_{;N} \ \text{(when } n=3\text{)},$$

$$f_{;N}F_{;NN} \ , \ f_{;NN}F_{;N} \ , \ f_{;N}F_{;N}(\text{tr}K) \ , \ f_{;N}F_{;N}S \ \text{(when } n=4\text{)}.$$

The explicit form of these equations is (Branson and Gilkey 1990)

$$-4c_5-5c_6+\frac{3}{2}c_9=0, \quad (5.3.72)$$

$$-5e_6-4e_7+2e_{10}=0, \quad (5.3.73)$$

$$2e_1-10e_2+5e_3-2e_7=0, \quad (5.3.74)$$

$$-2e_1+10e_2-e_3-10e_4-2e_5-5e_6+6e_7+2e_8=0, \quad (5.3.75)$$

$$-5e_8+4e_9-5e_{10}=0. \quad (5.3.76)$$

Last, one has to use the Lemma expressed by Eq. (5.3.23) when $n=2,3,4$, bearing in mind that

$$E_1 \equiv E(D_1)=-b_x-b^2, \quad (5.3.77)$$

$$E_2 \equiv E(D_2)=b_x-b^2. \quad (5.3.78)$$

For example, when $n=2$, one finds

$$a_1(f,D_1)-a_1(f,D_2)-a_0\Big(f_{xx}+2bf_x,D_1\Big)=0, \quad (5.3.79)$$

which implies (with Dirichlet conditions for D_1 and Robin conditions for D_2)

$$\int_M \Big[6f(E_1-E_2)-6f_{xx}-12Sf_x\Big]+\int_{\partial M}\Big[-b_2fS-(3+b_1)f_{;N}\Big]=0. \quad (5.3.80)$$

The integrand of the interior term in Eq. (5.3.80) may be re-expressed as a total divergence, and hence one gets

$$\int_{\partial M}\Big[(12-b_2)fS+(3-b_1)f_{;N}\Big]=0, \quad (5.3.81)$$

which leads to

$$b_1 = 3 \ , \ b_2 = 12. \tag{5.3.82}$$

Further details concern only the repeated application of all these algorithms, and hence we refer the reader to Branson and Gilkey (1990). We should emphasize, however, that no proof exists, so far, that functorial methods lead to the complete calculation of *all* heat-kernel coefficients. For the time being one can only say that, when Dirichlet or Robin boundary conditions, or a mixture of the two (see section 5.4) are imposed, functorial methods have been completely successful up to the evaluation of the $a_{5/2}$ coefficient (see Kirsten (1998) and references therein).

5.4 Mixed boundary conditions

Mixed boundary conditions are found to occur naturally in the theory of fermionic fields, gauge fields and gravitation, in that some components of the field obey one set of boundary conditions, and the remaining part of the field obeys a complementary set of boundary conditions (Avramidi and Esposito 1997, Esposito *et al.* 1997). Here, we focus on some mathematical aspects of the problem, whereas more difficult problems are studied in chapter 6. The framework of our investigation consists, as in section 5.3, of a compact Riemannian manifold, say M, with smooth boundary ∂M. Given a vector bundle V over M, we assume that V can be decomposed as the direct sum

$$V = V_n \oplus V_d, \tag{5.4.1}$$

near ∂M. The corresponding projection operators are denoted by Π_n and Π_d, respectively. On V_n one takes Neumann boundary conditions modified by some endomorphism, say S, of V_n (see (5.2.16)), while Dirichlet boundary conditions hold on V_d. The (total) boundary operator reads therefore (Gilkey 1995)

$$\mathcal{B}f \equiv \Big[\Big(\Pi_n f \Big)_{;N} + S \Pi_n f \Big]_{\partial M} \oplus \Big[\Pi_d f \Big]_{\partial M}. \tag{5.4.2}$$

On defining

$$\psi \equiv \Pi_n - \Pi_d, \tag{5.4.3}$$

seven new universal constants are found to contribute to heat-kernel asymptotics for an operator of Laplace type, say P. In other words, the linear combination of projectors considered in Eq. (5.4.3) gives rise to seven new invariants in the calculation of heat-kernel coefficients up to a_2: one invariant contributes to $\tilde{a}_{3/2}(f,P,\mathcal{B})$, whereas the other six contribute to $\tilde{a}_2(f,P,\mathcal{B})$ (of course, the number of invariants is continuously increasing as one considers higher-order heat-kernel coefficients). The dependence on the boundary operator is emphasized by including it explicitly into the arguments of heat-kernel coefficients. One can thus write the general formulae (cf. (5.2.20) and (5.2.21))

$$\tilde{a}_{3/2}(f,P,\mathcal{B}) = \frac{\delta}{96}(4\pi)^{1/2}\int_{\partial M} \mathrm{Tr}\Big[\beta_1 f \psi_{|i}\, \psi^{|i}\Big] + a_{3/2}(f,P,\mathcal{B}), \tag{5.4.4}$$

$$\begin{aligned}\tilde{a}_2(f,P,\mathcal{B}) = \frac{1}{360}\int_{\partial M} \mathrm{Tr}\Big[&\beta_2 f \psi \psi_{|i}\, \Omega^{i}{}_N + \beta_3 f \psi_{|i}\, \psi^{|i} (\mathrm{tr}K) \\ &+ \beta_4 f \psi_{|i}\, \psi_{|j} K^{ij} + \beta_5 f \psi_{|i}\, \psi^{|i} S \\ &+ \beta_6 f_{;N} \psi_{|i}\, \psi^{|i} + \beta_7 f \psi_{|i}\, \Omega^{i}{}_N\Big] + a_2(f,P,\mathcal{B}),\end{aligned} \tag{5.4.5}$$

where $a_{3/2}(f,P,\mathcal{B})$ is formally analogous to Eq. (5.2.20), but with some universal constants replaced by linear functions of ψ, Π_n, Π_d, and similarly for $a_2(f,P,\mathcal{B})$ and Eq. (5.2.21). The work in Branson and Gilkey (1992b), Vassilevich (1995a) and Gilkey (1995) has fixed the following values for the universal constants $\{\beta_i\}$ occurring in Eqs. (5.4.4) and (5.4.5):

$$\beta_1 = -12,\ \beta_2 = 60,\ \beta_3 = -12,\ \beta_4 = -24, \tag{5.4.6}$$

$$\beta_5 = -120,\ \beta_6 = -18,\ \beta_7 = 0. \tag{5.4.7}$$

To obtain this result, it is crucial to bear in mind that the correct functorial formula for the endomorphism S is

$$\left[\frac{d}{d\varepsilon} S(\varepsilon)\right]_{\varepsilon=0} = -fS + \frac{1}{2}(m-2) f_{;N} \Pi_n. \tag{5.4.8}$$

This result was first obtained in Vassilevich 1995a, where the author pointed out that Π_n should be included, since the variation of S should compensate the one of ω_n only on the subspace V_n. The unfortunate omission of Π_n led to incorrect results in physical applications, which were later corrected in Moss and Poletti (1994), hence confirming the analytic results in D'Eath and Esposito (1991a) and Esposito *et al.* (1994a).

There is indeed a very rich literature on the topics in spectral geometry studied in this chapter and in the previous one. In particular, further to the literature cited so far, we would like to recommend the work in Gilkey (1975), Kennedy (1978), Goldthorpe (1980), Widom (1980), Obukhov (1982,1983), Bérard (1986), Cognola and Zerbini (1988), Amsterdamski *et al.* (1989), Avramidi (1989), Moss and Dowker (1989), Avramidi (1990a,b, 1991), Gusynin *et al.* (1991), McAvity and Osborn (1991a,b), McAvity (1992), Barvinsky *et al.* (1994), Fursaev (1994), Dowker and Apps (1995), Endo (1995), Fulling (1995), Alexandrov and Vassilevich (1996), Avramidi and Schimming (1996). Yet other outstanding work can be found in Bellisai (1996), Bordag *et al.* (1996a–c), Dowker (1996a,b), Dowker *et al.* (1996), Dowker and Kirsten (1996), Falomir *et al.* (1996a,b), Yajima (1996), Avramidi (1997b), De Nardo *et al.* (1997), Estrada and Fulling (1997), Falomir (1997), Fursaev and Miele (1997), van de Ven (1997), Yajima (1997), Levitin (1998).

5.5 Heat equation and index theorem

Heat-kernel asymptotics makes it possible to obtain a deep and elegant formula for the index of elliptic operators. The problem has been studied by many authors (e.g. Atiyah and Patodi 1973, Günther and Schimming 1977, Gilkey 1995). Here, we follow the outline given in Atiyah (1975a).

Let P be an elliptic differential operator on a compact manifold M without boundary, and let $P^{\dagger}$ be its adjoint. One can then consider the two self-adjoint operators $P^{\dagger}P$ and $PP^{\dagger}$. If ϕ is any eigenfunction of $P^{\dagger}P$ with eigenvalue λ:

$$P^{\dagger}P\,\phi = \lambda\,\phi, \tag{5.5.1}$$

one has, acting with P on the left:

$$(PP^{\dagger})P\,\phi = \lambda\,P\phi. \tag{5.5.2}$$

This means that $P\phi$ is an eigenfunction of $PP^{\dagger}$ with eigenvalue λ, provided that λ does not vanish. Conversely, if Φ is any eigenfunction of $PP^{\dagger}$ with eigenvalue λ:

$$PP^{\dagger}\,\Phi = \lambda\,\Phi, \tag{5.5.3}$$

one finds

$$(P^\dagger P)P^\dagger\,\Phi = \lambda\,P^\dagger\Phi, \tag{5.5.4}$$

i.e. $P^\dagger\Phi$ is an eigenfunction of $P^\dagger P$ with eigenvalue λ, provided that $\lambda \neq 0$. In other words, the operators $P^\dagger P$ and $PP^\dagger$ have the same non-zero eigenvalues.

At this stage, one has to consider the operators $e^{-tP^\dagger P}$ and $e^{-tPP^\dagger}$. They are fundamental solutions of the corresponding heat equations (cf. Eq. (5.2.2)). For positive values of t, these have C^∞ kernels and hence are of trace class. Their L^2 traces read

$$\mathrm{Tr}_{L^2}\left(e^{-tP^\dagger P}\right) = \sum_\lambda e^{-t\lambda}, \tag{5.5.5}$$

$$\mathrm{Tr}_{L^2}\left(e^{-tPP^\dagger}\right) = \sum_\mu e^{-t\mu}, \tag{5.5.6}$$

where the summations run over the respective spectra. Thus, bearing in mind that the non-vanishing λ coincide with the non-vanishing μ, while $\lambda = 0$ corresponds to the null-space of P, and $\mu = 0$ corresponds to the null-space of $P^\dagger$, one finds

$$\mathrm{index}(P) = \mathrm{Tr}_{L^2}\left(e^{-tP^\dagger P}\right) - \mathrm{Tr}_{L^2}\left(e^{-tPP^\dagger}\right). \tag{5.5.7}$$

On the other hand, it is well known from section 5.2, following the work in Minakshisundaram and Pleijel (1949) and Greiner (1971), that for any elliptic self-adjoint differential operator A of order n with non-negative spectrum, the integrated heat kernel has an asymptotic expansion (m being the dimension of M)

$$\mathrm{Tr}_{L^2} e^{-tA} \sim \sum_{k=-m}^{\infty} a_k\, t^{k/n}. \tag{5.5.8}$$

The coefficients a_k are obtained by integrating over M a linear combination of local invariants (see (5.2.17)–(5.2.21)):

$$a_k = \int_M \alpha_k. \tag{5.5.9}$$

In our case, Eqs. (5.5.7)–(5.5.9) lead to a formula of the kind

$$\mathrm{index}(P) = \int_M (\alpha_0 - \beta_0), \tag{5.5.10}$$

where α_0 and β_0 refer to the operators $P^\dagger P$ and $PP^\dagger$, respectively. If the manifold M has a boundary, boundary terms contribute to the index as well. As we know from sections 3.6 and 3.7, it is sufficient to refer to the index of the Dirac operator, say $\mathcal{D}$, which takes the general form

$$\text{index}(\mathcal{D}) = \int_M (\alpha_0 - \beta_0) + \int_{\partial M} (\gamma_0 - \delta_0), \tag{5.5.11}$$

where γ_0 and δ_0 are linear combinations of local invariants on the boundary.

5.6 Heat kernel and ζ-function

If $\mathcal{A}$ is an elliptic, self-adjoint, positive-definite operator, the spectral theorem makes it possible to define its complex power $\mathcal{A}^{-s}$, where s is any complex number (Seeley 1967, Esposito *et al.* 1997). The L^2 trace of the complex power of $\mathcal{A}$ is, by definition, its ζ-function:

$$\zeta_{\mathcal{A}}(s) \equiv \mathrm{Tr}_{L^2} \mathcal{A}^{-s}. \tag{5.6.1a}$$

In the literature, Eq. (5.6.1a) is frequently re-expressed in the form

$$\zeta_{\mathcal{A}}(s) = \sum_{\lambda > 0} \lambda^{-s}, \tag{5.6.1b}$$

where λ runs over the (discrete) eigenvalues of $\mathcal{A}$, counted with their multiplicity. As it stands, this infinite sum converges if $\mathrm{Re}(s)$ is greater than a lower limit depending on the dimension m of the Riemannian manifold under consideration, and on the order n of the operator $\mathcal{A}$. Interestingly, $\zeta_{\mathcal{A}}(s)$ can be analytically continued as a meromorphic function on the whole s-plane (Seeley 1967, Atiyah *et al.* 1976, Elizalde 1995, Esposito *et al.* 1997), which is regular at $s = 0$. Moreover, a deep link exists between the ζ-function of $\mathcal{A}$ and its integrated heat kernel. This is expressed by the so-called inverse Mellin transform (Hawking 1977, Gilkey 1995):

$$\zeta_{\mathcal{A}}(s) = \frac{1}{\Gamma(s)} \int_0^\infty t^{s-1} \, \mathrm{Tr}_{L^2} \left(e^{-t\mathcal{A}} \right) dt, \tag{5.6.2}$$

where Γ is the Γ-function, defined in (4.2.28). If $\mathcal{A}$ is a second-order operator, the identity (5.6.2), jointly with the asymptotic expansion

$$\mathrm{Tr}_{L^2} \left(e^{-t\mathcal{A}} \right) \sim \sum_{k=0}^{\infty} A_{k/2} t^{(k-m)/2}, \tag{5.6.3}$$

implies that $\zeta(0)$ takes the form

$$\zeta_{\mathcal{A}}(0) = A_{m/2} = \int_M Q_{m/2} + \int_{\partial M} S_{m/2}, \qquad (5.6.4)$$

where $Q_{m/2}$ and $S_{m/2}$ are linear combinations of geometric invariants. Their expression can be derived, for a given value of m, from the general method described in sections 5.2 and 5.3. In quantum field theory, Eq. (5.6.4) is found to express the conformal anomaly or the one-loop divergence (Hawking 1977, Esposito *et al.* 1997).

Moreover, for a given elliptic operator, say D, one can define the operators (cf. section 5.5)

$$\Delta_0 \equiv 1 + D^\dagger D, \qquad (5.6.5)$$

$$\Delta_1 \equiv 1 + D D^\dagger, \qquad (5.6.6)$$

and the ζ-functions

$$\zeta_i(s) \equiv \mathrm{Tr}_{L^2} \Delta_i^{-s} \quad i = 1, 2. \qquad (5.6.7)$$

It is then possible to prove that (Atiyah 1966)

$$\mathrm{index}(D) = \zeta_0(s) - \zeta_1(s). \qquad (5.6.8)$$

Remarkably, this formula holds *for all values* of s. In particular, it is more convenient to choose $s = 0$ for the explicit calculation. The equations (5.5.11), (5.6.4) and (5.6.8) show the deep link between index theory, heat kernels and ζ-functions.

5.7 Majorizations of the heat kernel

The methods of analysis and geometry lead to majorizations of the heat kernel. These have been studied not only for mathematical completeness (Davies 1989), but also to derive the form of the first heat-kernel coefficients for manifolds with boundary. Following the seminal paper by McKean and Singer (1967), we now describe the basic elements of such an investigation, to complete our review of some powerful methods for the derivation of the explicit form of heat-kernel asymptotics.

Let M be a closed, m-dimensional, smooth manifold and let $Q : C^\infty(M) \rightarrow C^\infty(M)$ be a second-order elliptic operator which, on a coordinate patch $U \subset M$, can be written as (for simplicity, no emphasis is here put on the vector bundles over M, in agreement with the style of the original paper)

$$Q = -a^{ij}(x)\frac{\partial^2}{\partial x^i \partial x^j} + b^i(x)\frac{\partial}{\partial x^i}, \tag{5.7.1}$$

where the coefficients a^{ij} and b^i are smooth functions on U, and the quadratic form based upon the a^{ij} is positive-definite:

$$\sum_{i,j} a^{ij} y_i y_j > 0 \quad \forall y \neq 0.$$

The operator Q can be re-expressed as the sum of the Laplace–Beltrami operator (we insert a minus sign to obtain a positive spectrum, in agreement with our previous convention)

$$\triangle \equiv -\frac{1}{\sqrt{\det g}}\frac{\partial}{\partial x^i} g^{ij}\sqrt{\det g}\frac{\partial}{\partial x^j},$$

and a first-order differential operator, so that

$$Q = \triangle + h^i \frac{\partial}{\partial x^i}. \tag{5.7.2}$$

The operator $\triangle$ is symmetric, in that

$$\int u \triangle v \sqrt{\det g}\, dx = \int v \triangle u \sqrt{\det g}\, dx$$

$\forall u, v$ in its domain, and non-negative:

$$\int u \triangle u \sqrt{\det g}\, dx \geq 0$$

with respect to the Riemannian volume element $\sqrt{\det g}\, dx$. The same holds with respect to the volume element $e^w \sqrt{\det g}\, dx$, provided that the one-form dual to the vector field $h^i \frac{\partial}{\partial x^i}$ is an exact differential, equal to dw (section 2 of McKean and Singer (1967)). Consider now the heat kernel $G(x, y; t)$ of the heat equation for the operator Q,

$$\left(\frac{\partial}{\partial t} + Q\right) u = 0,$$

with respect to the volume element $\sqrt{\det g}\, dx$. The work by Minakshisundaram (1953) ensures that (cf. section 5.2)

$$0 < G(x, y; t) \in C^\infty[(0, \infty) \times M^2], \tag{5.7.3}$$

$$\left(\frac{\partial}{\partial t} + Q_x\right) G(x, y; t) = 0, \tag{5.7.4a}$$

$$\left(\frac{\partial}{\partial t}+Q_y^*\right)G(x,y;t)=0, \tag{5.7.4b}$$

$$\lim_{t\to 0}\int_M G(x,y;t)\sqrt{\det g}\,dy=1. \tag{5.7.5}$$

With this notation, Q^* is the dual of Q with respect to $\sqrt{\det g}\,dx$. Moreover, according to a useful result of Varadhan (1967),

$$\lim_{t\to 0}\frac{1}{t}\log(G(x,y;t))=-\frac{1}{4}d_R(x,y), \tag{5.7.6}$$

where $d_R(x,y)$ is the Riemannian distance between x and y. If Q is symmetric with respect to the volume element $e^w\sqrt{\det g}\,dx$, then $G(x,y;t)e^{-w(y)}$ is symmetric in x and y, and since the trace of the heat-kernel diagonal:

$$G(t)\equiv\int_M G(x,x;t)e^w\sqrt{\det g}\,dx \tag{5.7.7}$$

converges, the map

$$e^{-tQ}:f\to\int_M G(x,x;t)fe^w\sqrt{\det g}\,dx$$

is a compact map of the Hilbert space $\mathcal{H}\equiv L^2\Big[M,e^w\sqrt{\det g}\,dx\Big]$. Thus, the operator Q has a discrete spectrum (cf. Chavel 1984)

$$0=\lambda_0<\lambda_1\le\lambda_2\le\ \dots$$

with corresponding eigenfunctions $f_n\in C^\infty(M)$ which form a basis of $\mathcal{H}$. Moreover, the heat kernel is given by

$$G(x,y;t)=\sum_{n=0}^{\infty}f_n(x)\otimes f_n(y)e^{-\lambda_n t}, \tag{5.7.8}$$

where the right-hand side converges uniformly on compact figures of $(0,\infty)\times M^2$, and the *integrated heat kernel* (5.7.7) is (cf. (5.1.20))

$$G(t)=\int_M\sum_{n=0}^{\infty}e^{-\lambda_n t}f_n^2(x)e^w\sqrt{\det g}dx=\sum_{n=0}^{\infty}e^{-\lambda_n t}. \tag{5.7.9}$$

The idea of Kac (1966), and then used by McKean and Singer (1967), was to make first a *pointwise estimate* of the pole given by the heat-kernel diagonal $G(x,x;t)$, and then to integrate over M to get an estimate of $G(t)$.

Consider now a little closed patch U of M with smooth $(m-1)$-dimensional boundary B. If U is viewed as a part of $\Re^m$, one can extend $Q' \equiv Q \mid U$ to the whole of $\Re^m$ in such a way that the coefficients of the extension are C^∞ functions on $\Re^m$, and

$$Q' = -\sum_{i=1}^{m} \frac{\partial^2}{\partial x_i^2} \tag{$*$}$$

near ∞. Denoting by G' the heat kernel of the heat equation for the operator Q', we are now going to prove, following McKean and Singer (1967), that inside $U \times U$ the following inequality holds as $t \rightarrow 0^+$:

$$\mid G'(x,y;t) - G(x,y;t) \mid \leq e^{-\frac{\kappa}{t}}, \tag{5.7.10}$$

where κ is a positive constant depending only on the distance to B. Indeed, let G'' be the heat kernel of the heat equation

$$\left(\frac{\partial}{\partial t} + Q\right) u = 0, \tag{5.7.11}$$

with the boundary condition

$$[u]_B = 0. \tag{5.7.12}$$

If v is a compact function $\in C^\infty(U)$, one can define the function

$$u \equiv \int_M \Big(G''(x,y;t) - G(x,y;t)\Big) v(y) \sqrt{\det g}\, dy, \tag{5.7.13}$$

which solves the heat equation for the operator Q on $(0,\infty) \times U$, and tends to 0 uniformly on $\overline{U}$, as $t \rightarrow 0^+$. This means that, in the set $[0,t] \times \overline{U}$, the absolute value of u is peaked on $[0,t] \times B$, so that the Varadhan estimate (5.7.6) can be applied to find the inequality

$$\mid u \mid \leq \max_{[0,t]\times B} \mid \int_M (G'' - G) v \mid \; \leq e^{-\frac{R_s^2}{5t}} \|v\|_1. \tag{5.7.14}$$

With this notation, R_s is the shortest Riemannian distance from $(v \neq 0) \subset U$ to B.

By virtue of (5.7.10) it is possible, in the course of estimating $G(x,x;t)$ up to an exponentially small error, to replace M by the flat m-dimensional space $\Re^m$, and to assume that Q reduces to the form ($*$) far outside. The next step of the proof is the introduction of the operator Q^0 obtained from Q by *freezing* its coefficients at $y \in \Re^m$, with corresponding heat kernel $G^0(x,y;t)$:

$$G^0(x,y;t) = (4\pi t)^{-\frac{m}{2}} e^{-\frac{|\sqrt{a_0}(y-x-b_0)|^2}{4t}}. \tag{5.7.15}$$

Bearing in mind the properties (5.7.4)–(5.7.6), one can express the difference of heat kernels, $G - G^0$, in the form

$$G(x,y;t) - G^0(x,y;t) = \int_0^t ds \frac{\partial}{\partial s} \int_{\Re^m} G(x,\cdot\,;s) G^0(\cdot\,,y;t-s)$$

$$= \int_0^t ds \int_{\Re^m} \left(G^0 Q^* G - G Q^0 G^0 \right)$$

$$= \int_0^t ds \int_{\Re^m} G(x,\cdot\,;s)(Q - Q^0) G^0(\cdot\,,y;t-s). \tag{5.7.16}$$

Interestingly, this property can be written in the form

$$G = G^0 + G \cdot f, \tag{5.7.17}$$

where the symbol $\cdot$ denotes the *composition* on the last line of (5.7.16), and we have defined

$$f \equiv (Q - Q^0) G^0(x,y;t-s). \tag{5.7.18}$$

The repeated application of Eq. (5.7.17):

$$G = G^0 + G \cdot f = G^0 + (G^0 + G \cdot f) \cdot f = ...,$$

leads to a formal n-fold sum for the heat kernel $G(x,y;t)$, i.e.

$$G = G^0 + \sum_{n \geq 1} G^0 \cdot f \cdot ... \cdot f, \tag{5.7.19}$$

which is called the Levi sum. Interestingly, this formal sum turns out to converge uniformly to G on the sets $(0,\infty) \times \Re^{2m}$. To prove it, note that, since the operator Q is assumed to reduce to the flat Laplacian as one approaches infinity, the function defined in (5.7.18) satisfies the inequality

$$| f | \leq C_1 \left(\frac{| x - y |^3}{t^2} + \frac{| x - y |}{t} + 1 \right) t^{-\frac{m}{2}} e^{-C_2 \frac{|x-y|^2}{t}}$$

$$\leq C_3 t^{-\frac{(m+1)}{2}} e^{-C_4 \frac{|x-y|^2}{t}}, \tag{5.7.20}$$

where C_1, C_2, C_3 and C_4 are some positive constants. This result leads, in turn, to the inequality

$$| G^0 \cdot f \cdot ... \cdot f | \leq (C_5)^n [(n/2)!]^{-1} t^{\frac{(n-m)}{2}} e^{-C_6 \frac{|x-y|^2}{t}}, \tag{5.7.21}$$

where C_5 and C_6 are another pair of positive constants. Thus, the joint effect of (5.7.15) and (5.7.21) is the desired inequality

$$| G(x,y;t) | \leq \sum_{n=0}^{\infty} \frac{(C_5 \sqrt{t})^n}{(n/2)!} t^{-\frac{m}{2}} e^{-C_6 \frac{|x-y|^2}{t}}. \tag{5.7.22}$$

By construction, this $G(x,y;t)$ is the heat kernel of the heat equation (5.7.11). Such an equation has only one kernel subject to the inequality (5.7.22), and hence the formula (5.7.19) represents the desired solution. In fact, any kernel subject to (5.7.22) is also a solution of Eq. (5.7.17), and one can prove that the solution of (5.7.17) and (5.7.22) exists and is unique.

Levi's sum (5.7.19) was used by McKean and Singer to obtain the asymptotic expansion

$$(4\pi t)^{\frac{m}{2}} G(x,x;t) \sim 1 + \frac{t}{3}R - \frac{t}{2}\mathrm{div}h - \frac{t}{4} \mid h \mid^2 + \mathrm{O}(t^2), \quad (5.7.23)$$

where R is the trace of the Ricci tensor of the background, $\mathrm{div}h$ is the divergence

$$\mathrm{div}h \equiv \frac{1}{\sqrt{\det g}} \frac{\partial}{\partial x^i} \left(h^i \sqrt{\det g} \right),$$

and $\mid h \mid$ is the Riemannian length

$$\mid h \mid \equiv g(h,h) = g_{ij} h^i h^j .$$

This result was the first step towards their analysis of manifolds with boundary, where they obtained the coefficient of t^2 in the heat-kernel asymptotics. The method of McKean and Singer is not merely of historical interest, because it still provides a useful way of studying boundary-value problems where the boundary operator is very complicated (cf. Avramidi and Esposito 1998a, and our section 6.6).

Appendix 5.A

The original paper by Kac (1966) is so enlightening and enjoyable that a more detailed description of its content seems quite appropriate in a book with pedagogical aims like the present monograph. Kac divided it into 15 short sections, and the points we have selected for the general reader are as follows.

If a membrane Ω, held fixed along its boundary Γ, is set in motion, one finds that its displacement $F(x,y;t) \equiv F(\vec{\rho};t)$ in the direction perpendicular to its original plane obeys the wave equation (remember that, with our conventions, the second derivatives in the Laplacian are weighted with negative signs)

$$\frac{\partial^2 F}{\partial t^2} = -\kappa^2 \triangle F, \quad (5.A.1)$$

where the constant κ depends on the physical properties of the membrane as well as on the tension under which the membrane is held. Interestingly, solutions exist which are harmonic in time, i.e.

$$F(\vec{\rho}, t) = U(\vec{\rho})e^{i\omega t}, \tag{5.A.2}$$

and hence represent the *pure tones* (or *normal modes*) that the membrane is able to produce. Upon insertion into the wave equation, the solution (5.A.2) leads to the eigenvalue equation for the Laplacian,

$$\Delta U = \frac{\omega^2}{\kappa^2} U, \tag{5.A.3}$$

subject to the Dirichlet boundary condition

$$U = 0 \text{ on } \Gamma. \tag{5.A.4}$$

The meaning of (5.A.4) is that, *for a sufficiently smooth boundary*, $U(\vec{\rho})$ tends to 0 as $\vec{\rho}$ approaches a point of Γ from the inside. The theory of integral equations led eventually to the proof that a membrane has a *discrete spectrum of pure tones*. Thus, there exist (normalized) eigenfunctions, say ψ_n, of the operator $\kappa^2 \Delta$, which obey the eigenvalue equation

$$\kappa^2 \, \Delta \, \psi_n = \lambda_n \psi_n, \tag{5.A.5}$$

where $\lambda_1 \leq \lambda_2 \leq ...$, and are such that

$$\psi_n(\vec{\rho}) \rightarrow 0 \text{ as } \vec{\rho} \rightarrow \text{a point of } \Gamma. \tag{5.A.6}$$

The inverse problem in the theory of vibrating membranes studies to which extent the spectrum of the Laplace operator determines completely the shape of the vibrating body. More precisely, if Ω_1 and Ω_2 are two plane regions bounded by the curves Γ_1 and Γ_2, respectively, with boundary-value problems

$$\kappa^2 \, \Delta \, U = \lambda U \text{ in } \Omega_1, \tag{5.A.7}$$

$$U = 0 \text{ on } \Gamma_1, \tag{5.A.8}$$

and

$$\kappa^2 \, \Delta \, V = \mu V \text{ in } \Omega_2, \tag{5.A.9}$$

$$V = 0 \text{ on } \Gamma_2, \tag{5.A.10}$$

if, for each n, the eigenvalue λ_n for Ω_1 is equal to the eigenvalue μ_n for Ω_2, one would like to know whether the regions Ω_1 and

Ω_2 are congruent. Kac focused (as we do) on the asymptotic properties of large eigenvalues, i.e. (in the language of section 5.2) on the asymptotic expansion of the heat kernel $G(x,y;t)$ (the local asymptotics) and on the asymptotic form of the integrated heat kernel (the global asymptotics). Within this framework, the first non-trivial result is that one can 'hear the area' of Ω. Indeed, in the course of studying the standing electromagnetic waves within a cavity with a perfectly reflecting surface, the problem arises of proving that the number of sufficiently high overtones which lie in the frequency range $(\nu, \nu + d\nu)$ is independent of the shape of the cavity, and only depends on the volume through a linear relation. H. A. Lorentz conjectured in 1910 that the same holds for membranes, air masses and so on. In other words, denoting by $N(\lambda)$ the number of eigenvalues smaller than λ, one should find the asymptotic behaviour

$$N(\lambda) \sim \frac{|\,\Omega\,|}{2\pi}\lambda \text{ as } \lambda \rightarrow \infty. \tag{5.A.11}$$

The result (5.A.11) was proved by H. Weyl by using the theory of integral equations. Before we can derive it with the help of heat-equation methods, let us try to understand why the heat equation has been given this name. The physical problem is, of course, the one studied in *diffusion theory.* One should think that some sort of substance, initially concentrated at $\vec{\rho} \equiv (x_0, y_0)$, is diffusing through the plane region Ω bounded by Γ, and is absorbed at the boundary. Thus, the amount $P_\Omega(\vec{\rho}, \vec{r}; t)$ of matter at $\vec{r} = (x,y)$ at time t obeys the diffusion (or heat) equation

$$\frac{\partial P_\Omega}{\partial t} = -\kappa^2 \bigtriangleup P_\Omega, \tag{5.A.12}$$

jointly with the boundary condition

$$P_\Omega(\vec{\rho}, \vec{r}; t) \rightarrow 0 \text{ as } \vec{r} \rightarrow \text{ a boundary point}, \tag{5.A.13}$$

and the initial condition

$$\lim_{t\rightarrow 0} \int\int_A P_\Omega(\vec{\rho}, \vec{r}; t) d\vec{r} = 1 \tag{5.A.14}$$

for every open set A containing $\vec{\rho}$. The concentration $P_\Omega(\vec{\rho}, \vec{r}; t)$ can be expressed in terms of eigenvalues and eigenfunctions of the boundary-value problem (5.A.5) and (5.A.6). As we have written in section 5.1, this reads

$$P_\Omega(\vec{\rho},\vec{r};t) = \sum_{n=1}^{\infty} \psi_n(\vec{\rho})\psi_n(\vec{r})e^{-\lambda_n t}. \tag{5.A.15}$$

On the other hand, at small values of the parameter t, the particles of the diffusing material cannot yet 'feel' the presence of the boundary, and hence we expect that, as $t \to 0^+$, P_Ω has an asymptotic expansion expressed by $P_0(\vec{\rho},\vec{r};t)$, say. With this notation, P_0 is a solution for $t \to 0^+$ of the diffusion equation which is bounded from below and obeys the initial condition (5.A.14). Note that the property of being bounded from below is essential to ensure uniqueness of the solution, and that P_0 does not obey any boundary condition. The explicit calculation yields the well known result

$$P_0(\vec{\rho},\vec{r};t) = \frac{1}{2\pi t}\exp\left[-\frac{d_{r\rho}^2}{2t}\right], \tag{5.A.16}$$

where $d_{r\rho}$ is the Euclidean distance between $\vec{\rho}$ and $\vec{r}$.

According to the physical interpretation of the diffusion equation, one should have the inequality

$$P_\Omega(\vec{\rho},\vec{r};t) \leq P_0(\vec{\rho},\vec{r};t) = \frac{\exp\left[-\frac{d_{r\rho}^2}{2t}\right]}{2\pi t}, \tag{5.A.17}$$

because, if matter can get absorbed at the boundary Γ of Ω, less 'stuff' can be found at $\vec{r}$ at time t, with respect to the case when no absorption occurs. Let us now consider a square Q centred at $\vec{\rho}$ and totally contained in Ω. Its boundary is assumed to act as an absorbing barrier, and the corresponding concentration at $\vec{r}$ at time t is denoted by $P_Q(\vec{\rho},\vec{r};t)$, for $\vec{r} \in Q$. In this case, the Green's kernel is known explicitly, and the heat-kernel diagonal reads (a being the side of the square)

$$P_Q(\vec{\rho},\vec{\rho};t) = \frac{4}{a^2}\sum_{m,n(o)} \exp\left[-\frac{(m^2+n^2)\pi^2}{2a^2}t\right], \tag{5.A.18}$$

where $\sum_{m,n(o)}$ is our notation for the summation over the odd values of the integers m and n. Within the square Q, this is a solution for $t > 0$ of the equation (5.A.12) subject to the conditions (5.A.13) and (5.A.14). Since Q is totally contained in Ω, one can impose the inequalities

$$P_Q(\vec{\rho},\vec{r};t) \leq P_\Omega(\vec{\rho},\vec{r};t) \leq P_0(\vec{\rho},\vec{r};t), \tag{5.A.19}$$

which hold for all $\vec{r} \in Q$ and hence, in particular, when $\vec{r} = \vec{\rho}$. One then finds, from (5.A.15), (5.A.18) and (5.A.19),

$$\frac{4}{a^2} \sum_{m,n(o)} \exp\left[-\frac{(m^2+n^2)\pi^2}{2a^2}t\right] \leq \sum_{n=1}^{\infty} \psi_n^2(\vec{\rho})e^{-\lambda_n t} \leq \frac{1}{2\pi t}, \tag{5.A.20}$$

which implies that, as $t \rightarrow 0^+$, one has the asymptotic formula

$$\frac{4}{a^2} \sum_{m,n(o)} \exp\left[-\frac{(m^2+n^2)\pi^2}{2a^2}t\right] \sim \frac{1}{2\pi t}. \tag{5.A.21}$$

This leads, in turn, to the following asymptotic estimate as $t \rightarrow 0^+$:

$$\sum_{n=1}^{\infty} \psi_n^2(\vec{\rho})e^{-\lambda_n t} \sim \frac{1}{2\pi t}. \tag{5.A.22}$$

As a next step, we may integrate over Q the first of the inequalities (5.A.20), to obtain

$$4 \sum_{m,n(o)} \exp\left[-\frac{(m^2+n^2)\pi^2}{2a^2}t\right] \leq \sum_{n=1}^{\infty} e^{-\lambda_n t} \int\int_Q \psi_n^2(\vec{\rho})d\vec{\rho}. \tag{5.A.23}$$

One then covers Ω with a net of squares of side a, keeping only those contained in Ω. On denoting by $N(a)$ the number of these squares, and by $\Omega(a)$ the union of all these squares, one finds the inequality

$$\begin{aligned}\sum_{n=1}^{\infty} e^{-\lambda_n t} &= \sum_{n=1}^{\infty} e^{-\lambda_n t} \int\int_\Omega \psi_n^2(\vec{\rho})d\vec{\rho} \\ &\geq \sum_{n=1}^{\infty} e^{-\lambda_n t} \int\int_{\Omega(a)} \psi_n^2(\vec{\rho})d\vec{\rho} \\ &\geq 4N(a) \sum_{m,n(o)} \exp\left[-\frac{(m^2+n^2)\pi^2}{2a^2}t\right].\end{aligned} \tag{5.A.24}$$

Moreover, the second of the inequalities (5.A.20) may be integrated over Ω, to find

$$\sum_{n=1}^{\infty} e^{-\lambda_n t} \leq \frac{|\,\Omega\,|}{2\pi t}. \tag{5.A.25}$$

The results (5.A.24) and (5.A.25) may now be combined, and bearing in mind that $N(a)a^2 = \Omega(a)$, one eventually obtains

$$|\,\Omega(a)\,| \frac{4}{a^2} \sum_{m,n(o)} \exp\left[-\frac{(m^2+n^2)\pi^2}{2a^2} t\right] \leq \sum_{n=1}^{\infty} e^{-\lambda_n t}$$
$$\leq \frac{|\,\Omega\,|}{2\pi t}. \tag{5.A.26}$$

Moreover, the asymptotic property (5.A.21) implies that

$$\lim_{t \to 0} 2\pi t \frac{4}{a^2} \sum_{m,n(o)} \exp\left[-\frac{(m^2+n^2)\pi^2}{2a^2} t\right] = 1, \tag{5.A.27}$$

and this may be used to derive, from (5.A.26), another inequality:

$$|\,\Omega(a)\,| \leq \liminf_{t \to 0} 2\pi t \sum_{n=1}^{\infty} e^{-\lambda_n t}$$
$$\leq \limsup_{t \to 0} 2\pi t \sum_{n=1}^{\infty} e^{-\lambda_n t} \leq |\,\Omega\,|. \tag{5.A.28}$$

The analysis is completed if one remarks that, by choosing a sufficiently small size of the squares, one can make $|\,\Omega(a)\,|$ arbitrarily close to $|\,\Omega\,|$, so that (5.A.28) leads to

$$\lim_{t \to 0} 2\pi t \sum_{n=1}^{\infty} e^{-\lambda_n t} = |\,\Omega\,|, \tag{5.A.29}$$

which implies

$$\sum_{n=1}^{\infty} e^{-\lambda_n t} \sim \frac{|\,\Omega\,|}{2\pi t} \text{ as } t \to 0^+, \tag{5.A.30}$$

in complete agreement with Weyl's result. However, the derivation of (5.A.22) and (5.A.30) is not yet on solid ground, because a rigorous foundation for the inequalities (5.A.17) and (5.A.19) is still lacking. For this purpose, one has to combine statistical concepts with some rigorous measure theory.

To begin, one may view diffusion as a macroscopic manifestation of microscopic Brownian motion, according to the ideas of Einstein and Smoluchowski. In other words, $P_\Omega(\vec{\rho},\vec{r};t)$ is interpreted as the probability density of finding a free Brownian particle at $\vec{r}$ at time t, if it started on its journey at $t=0$ from $\vec{\rho}$, with the understanding that absorption of the particle occurs when it comes to the boundary. Thus, if a large number N of independent free Brownian particles start from $\vec{\rho}$, the integral

$$N\int\int_A P(\vec{\rho},\vec{r};t)d\vec{r}$$

represents the average number of these particles which are found in A at time t. The larger is N, the better is the description provided by continuous diffusion theory, because the statistical percentage error is of order $N^{-1/2}$.

One may, however, obtain a deeper understanding if, instead of thinking of the statistics of particles, one formulates the problem in terms of *statistics of paths*. This point of view is due to N. Wiener. Consider the set of all continuous curves $r(\tau), \tau \in [0,\infty[$, starting from some arbitrarily chosen origin O. Let $\Omega_1,\Omega_2,...,\Omega_n$ be open sets, and $t_1 < t_2 < ... < t_n$ some ordered instants of time. The theory of Einstein and Smoluchowski requires that the probabilities

$$\text{Prob}\,\{\rho + r(t_1) \in \Omega_1, \rho + r(t_2) \in \Omega_2, ..., \rho + r(t_n) \in \Omega_n\}$$

should be expressed by the integral

$$\int_{\Omega_1} \cdots \int_{\Omega_n} \prod_{j=0}^{n-1} P_0\Bigl(\vec{r}_j,\vec{r}_{j+1};t_{j+1}-t_j\Bigr)d\vec{r}_1...d\vec{r}_n,$$

where P_0 is given by (5.A.16), and $\vec{r}_0 \equiv \vec{\rho}, t_0 \equiv 0$. Interestingly, Wiener proved that it is possible to construct a completely additive measure on the space of all continuous curves $r(\tau)$ emanating from the origin, such that the set of curves $\rho+r(\tau)$, which at times $t_1 < t_2 < ... < t_n$ are in open sets $\Omega_1,...,\Omega_n$ respectively, has measure given by the Einstein–Smoluchowski formula just written.

What is very important for our purposes is that the set C_s of curves such that $\rho+r(\tau) \in \Omega, \tau \in [0,t]$, and $\rho+r(t)$ belongs to an open set, say A, is measurable, and if Ω has a sufficiently smooth boundary, one can also prove that this measure is equal to

$$\int_A P_\Omega(\vec{\rho},\vec{r};t)d\vec{r}. \qquad (*)$$

It is now clear why the inequalities (5.A.17) and (5.A.19) hold in a rigorous mathematical formulation. They simply express the property that, if the sets A_1, A_2, A_3 obey the inclusion relation

$$A_1 \subset A_2 \subset A_3,$$

the corresponding measures obey the inequalities

$$m(A_1) \leq m(A_2) \leq m(A_3). \tag{5.A.31}$$

It should be stressed that the set C_s of curves is measurable even if the boundary Ω is not smooth. The corresponding measure can still be written as the integral (*), and in the interior of Ω, $P_\Omega(\vec{\rho}, \vec{r}; t)$ satisfies the diffusion equation (5.A.12) and the initial condition (5.A.14), for all open sets A such that $\vec{\rho} \in A$. However, the boundary condition (5.A.13) does not seem to admit a clear interpretation under such circumstances. Thus, the classical theory of diffusion needs to assume that suitably regular boundaries exist.

It is now instructive to perform a comparison with some properties of the Schrödinger equation, which is the cornerstone of non-relativistic quantum theory. Suppose that the Hamiltonian operator, H, is independent of the time variable, is self-adjoint (more precisely, essentially self-adjoint on a suitable domain), and possesses a complete set of orthonormal eigenvectors, say $\{u_j\}$, and a purely discrete spectrum, say $\{E_j\}$. The the eigenvalue equation for H reads

$$Hu_j(\vec{x}) = E_j u_j(\vec{x}), \tag{5.A.32}$$

and the initial data for the Schrödinger equation can be expanded as

$$\psi(\vec{x}, 0) = \sum_{j=1}^{\infty} C_j u_j(\vec{x}), \tag{5.A.33}$$

where the coefficients C_j can be computed by the formula

$$C_j = \int u_j^*(\vec{x}) \psi(\vec{x}, 0) d^3x, \tag{5.A.34}$$

since eigenvectors belonging to different eigenvalues are orthogonal:

$$\int u_j^*(\vec{x}) u_l(\vec{x}) d^3x = \delta_{jl}.$$

The solution of the Schrödinger equation

$$i\hbar \frac{\partial \psi}{\partial t} = H\psi \tag{5.A.35}$$

is thus found to be

$$\psi(\vec{x},t) = e^{-itH/\hbar}\psi(\vec{x},0) = \sum_{j=1}^{\infty} C_j \sum_{r=0}^{\infty} \frac{(-it/\hbar)^r}{r!} H^r u_j(\vec{x})$$

$$= \sum_{j=1}^{\infty} C_j u_j(\vec{x}) e^{-iE_j t/\hbar}, \qquad (5.A.36)$$

where we have used the formal Taylor series for $e^{-itH/\hbar}$, jointly with the eigenvalue equation (5.A.32) and the purely discrete nature of the spectrum of H (when the Hamiltonian has also a continuous spectrum, the analysis is of course more elaborated, but quite standard by now (Fulling 1989)). In other words, the general solution is expressed as an infinite sum of elementary solutions $u_j(\vec{x})e^{-iE_j t/\hbar}$, and it is now clear why one first needs to solve the eigenvalue problem for the time-independent Schrödinger equation (5.A.32).

A very useful expression of the solution is obtained after inserting the result (5.A.34) for the coefficients into Eq. (5.A.36), which leads to

$$\psi(\vec{x},t) = \sum_{n=1}^{\infty} \int u_n(\vec{x}) u_n^*(\vec{x}') \psi(\vec{x}',0) e^{-iE_n t/\hbar} d^3x'$$

$$= \int G(\vec{x},\vec{x}';t)\psi(\vec{x}',0) d^3x', \qquad (5.A.37)$$

where $G(\vec{x},\vec{x}';t)$ is the standard notation for the Green's function (cf. (5.A.15)):

$$G(\vec{x},\vec{x}';t) \equiv \sum_{n=1}^{\infty} u_n(\vec{x}) u_n^*(\vec{x}') e^{-iE_n t/\hbar}. \qquad (5.A.38)$$

In other words, once the initial condition $\psi(\vec{x},0)$ is known, the solution at a time $t \neq 0$ is obtained by means of Eq. (5.A.37), where $G(\vec{x},\vec{x}';t)$ is called the Schrödinger kernel of the operator $e^{-itH/\hbar}$. This is, by definition, a solution for $t \neq 0$ of the equation

$$\left(i\hbar \frac{\partial}{\partial t} - H_{(x)}\right) G(\vec{x},\vec{x}';t) = 0, \qquad (5.A.39)$$

subject to the initial condition (φ being a suitably smooth function)

$$\lim_{t\to 0} \int G(\vec{x},\vec{x}';t)\varphi(\vec{x}') d^3x' = \varphi(\vec{x}). \qquad (5.A.40)$$

This is the correct mathematical way to express the distributional behaviour of the Schrödinger kernel. In the physics literature, Eqs. (5.A.39) and (5.A.40) are more frequently re-expressed as follows:

$$\left(i\hbar\frac{\partial}{\partial t} - H_{(x)}\right) G(\vec{x},\vec{x}';t) = \delta(\vec{x},\vec{x}')\delta(t), \tag{5.A.41}$$

$$G(\vec{x},\vec{x}';0) = \delta(\vec{x},\vec{x}'). \tag{5.A.42}$$

Similarly, in the case of the heat equation for the operator P (see (5.2.2)) the heat kernel $G(x,y;t)$ provides the solution in the form (Davies 1997)

$$U(\vec{x},t) = e^{-tP}f(\vec{x}) = \int_M G(x,y;t)f(y)\sqrt{\det g(y)}\,dy, \tag{5.A.43}$$

where $f(\vec{x})$ is the initial condition,

$$\lim_{t\rightarrow 0} U(\vec{x},t) = U(\vec{x},0) = f(\vec{x}). \tag{5.A.44}$$

In general, the various kernels have an integral representation

$$G(x,y;t) = \int_0^\infty \eta(t\lambda)dE_\lambda(x,y), \tag{5.A.45}$$

where E_λ is the spectral decomposition of the operator under consideration, and η is a smooth function on $(0,\infty)$. The methods of distribution theory and summability theory (Estrada and Kanwal 1990, 1994) show that the behaviour of the integrand at infinity determines whether the expansion of the Green function is genuinely asymptotic in the literal, pointwise sense, or is only valid in a distributional sense in the variable t. This is the difference between the heat kernel and the Schrödinger kernel (Estrada and Fulling 1997).

A recent series of exciting developments in the theory of heat-kernel asymptotics has been obtained on studying domains with a fractal boundary. One then deals with the *triadic von Koch snowflake* (Fleckinger *et al.* 1995). The mathematical problem begins with the analysis of the heat equation for the Laplace operator on an open set $D \subset \Re^n$. More precisely, denoting by

$$u(x,t) : D \times [0,\infty] \rightarrow \Re$$

the unique solution, for $t > 0$ and $x \in D$, of the heat equation

(our sign convention for the Laplacian $\triangle_D$ differs from the one used by Fleckinger *et al.*)

$$\left(\frac{\partial}{\partial t} + \triangle_D\right) u(x,t) = 0, \qquad (5.A.46)$$

subject to the condition $u(x,t) = 1$ for $x \in D$, the total amount of heat contained in D at $t \geq 0$ is

$$Q_D(t) = \int_D u(x,t)dx, \qquad (5.A.47)$$

and hence the total amount of heat lost up to the moment t is

$$E_D(t) = \int_D (1 - u(x,t))dx. \qquad (5.A.48)$$

Interestingly, if one studies the function E_D on an arbitrary open set in $\Re^n$ with a finite volume and a fractal boundary, one finds (assuming a uniform capacitary density condition) that a positive constant C exists such that (van den Berg 1994)

$$C^{-1}t^{\alpha} \leq E_D(t) \leq Ct^{\alpha}. \qquad (5.A.49)$$

In the inequality (5.A.49), one has

$$\alpha = 1 - \frac{d}{2}, \qquad (5.A.50)$$

where

$$d = 2\frac{\log(2)}{\log(3)} \qquad (5.A.51)$$

is called the Minkowski dimension of the boundary of the snow-flake.

The work by Fleckinger *et al.* (1995) has studied the asymptotic expansion as $t \rightarrow 0^+$ of $E_D(t)$, proving that, for the triadic von Koch snowflake, one has

$$E_D(t) \sim p(\log t)t^{\alpha} - q(\log t)t + \mathrm{O}(e^{-r/t}) \text{ as } t \rightarrow 0^+. \qquad (5.A.52)$$

The functions p and q are found to be continuous and $(\log 9)$-periodic, whereas r takes the peculiar value

$$r = \frac{1}{1152}. \qquad (5.A.53)$$

Remarkably, *only two* asymptotic terms occur, with an exponential remainder estimate. Thus, the fractal geometry of the boundary is another source of $\log t$ terms in the asymptotics as $t \rightarrow 0^+$, further to the case of non-local boundary conditions of the Atiyah–Patodi–Singer type (Grubb and Seeley 1995). We can only refer the reader to Fleckinger *et al.* (1995) for the detailed proof of (5.A.52), and for the beautiful pictures of the triadic von Koch snowflake, but we would like to point out that they begin with a careful application of the principle of 'not feeling the boundary' (see their Lemma 1.3), which was first introduced in the paper by Kac (1966) that we have discussed at length.

6
New frontiers

Quarks are spin-1/2 fields for which a Dirac operator can be studied. The local boundary conditions proposed for models of quark confinement are naturally related to the local boundary conditions studied, more recently, in one-loop quantum cosmology. Further developments lie in the possibility of studying quantization schemes in conformally invariant gauges. This possibility is investigated in the case of the Eastwood–Singer gauge for vacuum Maxwell theory on manifolds with boundary. This is part of a more general scheme, leading to the analysis of conformally covariant operators. These are also presented, with emphasis on the Paneitz operator. In spectral geometry, a class of boundary operators are described which include the effect of tangential derivatives. They lead to many new invariants in the heat-kernel asymptotics for operators of Laplace type. The consideration of tangential derivatives arises naturally within the framework of recent attempts to obtain Becchi–Rouet–Stora–Tyutin-invariant boundary conditions in quantum field theory. However, in Euclidean quantum gravity, it remains unclear how to write even just the general form of the various heat-kernel coefficients. Last, the role of the Dirac operator in the derivation of the Seiberg–Witten equations is described. The properties of the new scheme, with emphasis on the invariants and on the attempts to classify four-manifolds, are briefly introduced.

6.1 Introduction

So far we have dealt with many aspects of manifolds with boundary in mathematics and physics. Hence it seems appropriate to begin the last chapter of our monograph with a brief review of the areas of research which provide the main motivations for similar investigations. They are as follows.

(i) Index theorems for manifolds with boundary. One wants to understand how to extend index theorems, originally proved for closed manifolds, to manifolds with boundary (Atiyah and Singer 1963, Atiyah and Bott 1965, Atiyah 1975b). A naturally occurring analytic tool is indeed the use of complex powers of elliptic operators (Seeley 1967), via η- and ζ-functions.

(ii) Spectral geometry with mixed boundary conditions. Spinor fields, gauge fields and gravitation are subject to mixed boundary conditions. These may be local or non-local (Esposito *et al.* 1997), and reflect the first-order nature of the Dirac operator, and the symmetries of gauge fields and gravitation (e.g. invariance under local gauge transformations, invariance under (infinitesimal) diffeomorphisms, BRST symmetry, supersymmetry). The resulting boundary operators may be (complementary) projectors, or first-order differential operators. They lead to a heat-kernel asymptotics which is completely understood only in the cases involving a mixture of Dirichlet and Robin boundary conditions (Gilkey 1995, Esposito *et al.* 1997, Avramidi and Esposito 1998a,b).

(iii) Quantum cosmology. The path-integral approach to quantum cosmology needs a well-defined prescription for the 'sum over histories' which should define the quantum state of the universe (Hawking 1979, 1982, 1984, Hartle and Hawking 1983, Gibbons and Hawking 1993, Esposito 1994a, Moss 1996, Esposito *et al.* 1997).

(iv) Quantum field theory on manifolds with boundary. Boundaries play a crucial role in obtaining a complete and well defined mathematical model for the quantization of all fundamental interactions (Luckock 1991, Avramidi *et al.* 1996, Marachevsky and Vassilevich 1996, Avramidi and Esposito 1997, Esposito *et al.* 1997, Moss and Silva 1997, Vassilevich 1997). Indeed, one is already familiar with the need to specify boundary data from the analysis of potential theory (Kellogg 1954), waveguides, and the classical variational problem (cf. York 1986).

(v) Boundary counterterms in supergravity. The occurrence of boundaries provides a crucial test of the finiteness properties of supersymmetric theories of gravitation, e.g. simple supergravity (van Nieuwenhuizen and Vermaseren 1976, D'Eath 1996, Esposito and Kamenshchik 1996, Esposito 1996, Moniz 1996, Esposito *et al.* 1997).

(vi) Applications of quantum field theory. Variations of zero-point energies of the quantized electromagnetic field are finite, measurable, and depend on the geometry of the problem. This is what one learns from the Casimir effect (Casimir 1948, Sparnay 1958, Boyer 1968, Grib *et al.* 1994, Esposito *et al.* 1997, Lamoreaux 1997). Moreover, the consideration of boundary effects provides a useful toy model for the investigation of quark confinement, as is shown in the following section (Chodos *et al.* 1974).

6.2 Quark boundary conditions

A non-trivial application of spin-1/2 fields to modern physics consists of the mathematical models for quark fields. We here focus on the model proposed by Chodos *et al.* (1974), which describes elementary particles as composite systems, with their internal structure being associated with quark and gluon field variables. Their main idea was to account for the internal structure with the help of fields. However, the fields which describe the sub-structure of the hadron *belong only to the sub-structure of a particle*, and cannot be extended to all points of space, unlike the normal situation in field theory. In other words, field variables are confined on the subset of points which are inside an extended particle. Such a set of points represents *the bag*. Since then, the model has been called the M.I.T. bag model (the authors being research workers of the M.I.T.). Following the presentation in Johnson (1975), one can say that a local disturbance of some medium leads to a collective motion described by fields, which are taken to be the quark–gluon fields. The localized excitation consists of a particle, i.e. a hadron.

When the M.I.T. model was first proposed, there was (of course) great excitement in the physics community. Whether or not the model by Chodos *et al.* can still have an impact on current developments in particle physics, it provides a first relevant example of local boundary conditions on fields acted upon by a Dirac operator, and this is the feature that we want to emphasize hereafter. For this purpose, we follow again Johnson (1975), and we denote by $q_a(x)$ the quark field. This is a Dirac field carrying colour and flavour. Bearing in mind what we said at the beginning, the quark fields can only be associated with points *inside the hadron.* Outside the hadron, $q_a(x)$ vanishes by definition. The local flux of colour and flavour quantum numbers inside the hadron is

$$j_{ab}^{\mu}(x) \equiv \overline{q}_a(x)\, \gamma^{\mu}\, q_b(x). \tag{6.2.1}$$

To avoid losing quantum numbers through the surface, one has to impose the boundary condition (hereafter, n^{μ} is the inward-pointing normal to the boundary)

$$\Big[n_{\mu}\, j_{ab}^{\mu}(x)\Big]_{\partial M} = 0, \tag{6.2.2a}$$

where ∂M denotes the surface of the hadron. By virtue of (6.2.1), the boundary condition (6.2.2a) may be re-expressed in the form

$$\Big[\overline{q}_a(x)\, n_{\mu}\, \gamma^{\mu}\, q_b(x)\Big]_{\partial M} = 0. \tag{6.2.2b}$$

At this stage we remark that, denoting by ε a real-valued function of position on the boundary (ε was taken equal to ± 1 in Johnson (1975)):

$$\varepsilon : y \in \partial M \longrightarrow \varepsilon(y) \in \Re,$$

if one imposes the local boundary condition

$$\Big[\Big(i\, n_{\mu}\, \gamma^{\mu} - \varepsilon\Big) q_a(x)\Big]_{\partial M} = 0, \tag{6.2.3}$$

one obtains also the 'conjugate' boundary condition

$$\Big[\overline{q}_a(x)\Big(i\, n_{\mu}\, \gamma^{\mu} + \varepsilon\Big)\Big]_{\partial M} = 0. \tag{6.2.4}$$

By virtue of Eq. (6.2.3) one now finds

$$\Big[i\, n_{\mu}\, j_{ab}^{\mu}(x)\Big]_{\partial M} = \Big[\overline{q}_a(x)\, i\, n_{\mu}\, \gamma^{\mu}\, q_b(x)\Big]_{\partial M} = \Big[\varepsilon\, \overline{q}_a(x)\, q_b(x)\Big]_{\partial M}. \tag{6.2.5}$$

Moreover, by virtue of Eq. (6.2.4) one also has

$$\Big[i\, n_{\mu}\, j_{ab}^{\mu}(x)\Big]_{\partial M} = -\Big[\varepsilon\, \overline{q}_a(x)\, q_b(x)\Big]_{\partial M}. \tag{6.2.6}$$

The comparison of Eqs. (6.2.5) and (6.2.6) leads to

$$\Big[\overline{q}_a(x)\, q_b(x)\Big]_{\partial M} = 0, \tag{6.2.7}$$

which ensures the fulfillment of the condition (6.2.2a), bearing in mind the definition (6.2.1). Recent results on the bag model can be found in Wipf and Dürr (1995), Dürr and Wipf (1997).

The local boundary conditions (6.2.3) find another relevant physical application in a completely different framework, i.e. (one-loop) quantum cosmology, but one should bear in mind that the background of the bag model is three-dimensional, whereas quantum cosmological backgrounds studied so far in one-loop calculations are four-dimensional. A review of the necessary steps is given in the following section.

6.3 Quantum cosmology

In the early 1990s, there was an intensive investigation of local boundary conditions for the Dirac operator in one-loop quantum cosmology (e.g. D'Eath and Esposito 1991a, Kamenshchik and Mishakov 1993, Esposito 1994a, Moss and Poletti 1994). The motivations for this programme were as follows.

(i) If one studies local boundary conditions for solutions of the massless free-field equations, following the work of Breitenlohner and Freedman (1982) and Hawking (1983), one finds that there exist solutions of the Euclidean twistor equation which generate *rigid supersymmetry transformations* among *classical solutions* obeying boundary conditions of the type

$$\sqrt{2}\; {}_e n_A^{\ A'} \; \phi^A = \pm \widetilde{\phi}^{A'} \quad \text{on} \quad \partial M, \tag{6.3.1}$$

$$2\; {}_e n_A^{\ A'} \; {}_e n_B^{\ B'} \; \rho^{AB} = \pm \widetilde{\rho}^{A'B'} \quad \text{on} \quad \partial M. \tag{6.3.2}$$

(ii) Following instead the work in Luckock and Moss (1989) and Luckock (1991), one finds that *local supersymmetry transformations* exist in supergravity which fix at the boundary the spatial components of the tetrad and a projector acting on spinor-valued one-forms which represent the gravitino potentials. On extending this scheme to lower-spin fields in a supergravity multiplet, one obtains the local boundary conditions

$$\sqrt{2}\; {}_e n_A^{\ A'} \; \psi^A = \pm \widetilde{\psi}^{A'} \quad \text{on} \quad \partial M, \tag{6.3.3}$$

for a massless spin-1/2 field which can be expanded in a complete orthonormal set of eigenfunctions of the Dirac operator.

In particular, on studying a portion of flat Euclidean four-space bounded by a three-sphere of radius a (Schleich 1985), the boundary conditions (6.3.3) were found to imply the eigenvalue condition (D'Eath and Esposito 1991a, Esposito 1994a, and cf. Berry and Mondragon 1987)

$$F(E) \equiv [J_{n+1}(Ea)]^2 - [J_{n+2}(Ea)]^2 = 0 \quad \forall n \geq 0. \tag{6.3.4}$$

Although the boundary conditions (6.3.3) were a natural extension of the boundary conditions (6.2.3) to problems where the boundary and the background have an additional dimension, it was a non-trivial step, both technically and conceptually, to undertake the analysis of heat-kernel asymptotics with this class of local boundary conditions for the Dirac operator (see Falomir (1996) for the discussion of possible topological obstructions). Nevertheless, the result in D'Eath and Esposito (1991a; see also our appendix 6.A):

$$\zeta(0) = \frac{11}{360}, \tag{6.3.5}$$

which was later confirmed, independently, by several authors (Kamenshchik and Mishakov 1993, Moss and Poletti 1994, Dowker 1996a, Kirsten and Cognola 1996), provided evidence in favour of local boundary conditions for the Dirac operator being admissible in a number of relevant cases (the crucial point is to prove that local boundary conditions lead to ellipticity of the boundary-value problem; see Avramidi and Esposito 1997). Note that, by requiring that Eq. (6.3.3) should be preserved under the action of the Dirac operator, one obtains the complementary condition (Esposito 1994a)

$$\left[\left({}_e n^{BB'} \nabla_{BB'} + \frac{1}{2}(\mathrm{Tr}K)\right)\left(\sqrt{2}\, {}_e n_A^{\ A'}\, \psi^A \pm \widetilde{\psi}^{A'}\right)\right]_{\partial M} = 0. \tag{6.3.6}$$

In other words, denoting by P_- and P_+ two complementary projectors, where

$$P_{\pm} \equiv \frac{1}{2}\Big(1 \pm i\gamma_5\, \gamma_\mu\, n^\mu\Big), \tag{6.3.7}$$

the boundary conditions (6.3.3) and (6.3.6) can be cast in the form

$$\Big[P_- \phi\Big]_{\partial M} = 0, \tag{6.3.8}$$

$$\left[\left(n^a \nabla_a + \frac{1}{2}(\mathrm{Tr}K)\right) P_+ \phi\right]_{\partial M} = 0, \qquad (6.3.9)$$

which is the form of mixed boundary conditions studied in section 5.4.

Another important result was that, for a massless spin-1/2 field subject to non-local boundary conditions of the form described in section 4.4, one finds again the conformal anomaly (6.3.5) when the same flat background with boundary is studied. This result was obtained in D'Eath and Esposito (1991b) by using the Laplace transform of the heat equation and, independently, by Kamenshchik and Mishakov (1992), who used instead the powerful analytic technique derived by Barvinsky *et al.* (1992). The result was non-trivial because no geometric formulae for heat-kernel asymptotics with non-local boundary conditions were available at that time in the literature. Thus, analytic methods for $\zeta(0)$ calculations were the only way to investigate such issues (cf. Grubb 1996).

The current developments in one-loop quantum cosmology deal instead with mixed boundary conditions for gauge fields and gravitation, and with different quantization techniques for such fields. Since many topics have already been studied in detail by Esposito *et al.* (1994a,b, 1995a,b, 1997), we prefer to focus, in the following section, on open problems. As one might expect, the main source of fascinating open problems lies in the attempts to quantize the gravitational field.

6.4 Conformal gauges and manifolds with boundary

The recent attempts to quantize Euclidean Maxwell theory in quantum cosmological backgrounds have led to a detailed investigation of the quantized Maxwell field in covariant and non-covariant gauges on manifolds with boundary (Esposito 1994a,b, Esposito and Kamenshchik 1994, Vassilevich 1995a,b, Esposito *et al.* 1997). The main emphasis has been on the use of analytic or geometric techniques to evaluate the one-loop semiclassical approximation of the wave function of the universe, when magnetic or electric boundary conditions are imposed. In the former case one sets to zero at the boundary the tangential components A_k of the potential (the background value of A_μ is taken to vanish), the real-valued ghost fields ω and ψ (or, equivalently, a complex-valued ghost zero-form ε), and the gauge-averaging functional $\Phi(A)$ (see appendix 6.B):

$$\Big[A_k\Big]_{\partial M} = 0, \tag{6.4.1}$$

$$[\varepsilon]_{\partial M} = 0, \tag{6.4.2}$$

$$\Big[\Phi(A)\Big]_{\partial M} = 0. \tag{6.4.3}$$

In the electric scheme one sets instead to zero at the boundary the normal component of A_μ, jointly with the normal derivative of the ghost and the normal derivative of A_k:

$$\Big[A_0\Big]_{\partial M} = 0, \tag{6.4.4}$$

$$\Big[\partial \varepsilon / \partial n\Big]_{\partial M} = 0, \tag{6.4.5}$$

$$\Big[\partial A_k / \partial n\Big]_{\partial M} = 0. \tag{6.4.6}$$

The boundary conditions (6.4.1)–(6.4.3) and (6.4.4)–(6.4.6) are found to be invariant under infinitesimal gauge transformations on A_μ, as well as under BRST transformations (Esposito *et al.* 1997, Moss and Silva 1997).

On the other hand, the gauge-averaging functionals studied in Esposito (1994a,b), Esposito and Kamenshchik (1994), Vassilevich (1995a,b), Esposito *et al.* 1997, were not conformally invariant, although a conformally invariant choice of gauge was already known, at the *classical* level, from the work of Eastwood and Singer (1985). It has been therefore our aim to investigate the *quantum* counterpart of the conformally invariant scheme proposed in Eastwood and Singer (1985), to complete the current work on quantized gauge fields. For this purpose, we have studied a portion of flat Euclidean four-space bounded by three-dimensional surfaces. The vanishing curvature of the four-dimensional background is helpful to obtain a preliminary understanding of the quantum operators, which will be shown to have highly non-trivial properties. In our scheme, all curvature effects result from the boundary only (Esposito 1997b).

In flat Euclidean four-space, the conformally invariant gauge proposed in Eastwood and Singer (1985) reads (hereafter $b, c = 0, 1, 2, 3$)

$$\nabla_b \nabla^b \nabla^c A_c = \Box \nabla^c A_c = 0. \tag{6.4.7}$$

This supplementary condition is considered because the vacuum Maxwell equations in four dimensions are conformally invariant, and hence a scheme where both the field equations and the supplementary condition are conformally invariant appears quite interesting, if not desirable. If the classical potential is subject to an infinitesimal gauge transformation

$${}^f A_b = A_b + \nabla_b f, \tag{6.4.8}$$

the gauge condition (6.4.7) is satisfied by ${}^f A_b$ if and only if f obeys the fourth-order equation

$$\Box^2 f = 0, \tag{6.4.9}$$

where $\Box^2$ is the $\Box$ operator composed with itself, i.e. $\Box^2 \equiv g^{ab} g^{cd} \nabla_a \nabla_b \nabla_c \nabla_d$.

In the quantum theory via path integrals, however, one performs Gaussian averages over gauge functionals $\Phi(A)$ which ensure that well-defined Feynman Green's functions for the $\mathcal{P}$ operator on A_b, and for the ghost operator, actually exist (DeWitt 1981, Esposito *et al.* 1997). This means that the left-hand side of Eq. (6.4.7) is no longer set to zero. One defines instead a gauge-averaging functional

$$\Phi(A) \equiv \Box \nabla^b A_b, \tag{6.4.10}$$

and the gauge-averaging term $\frac{\beta}{2\alpha}[\Phi(A)]^2$, with β dimensionful and α a dimensionless parameter, is added to the Maxwell Lagrangian $\frac{1}{4} F_{ab} F^{ab}$. This is actually a crucial point which deserves further comments. If one considers the Lorenz functional $\Phi_L(A) \equiv \nabla^b A_b$, which only involves first derivatives of A_b, the resulting gauge-averaging term does not change the second-order nature of the operator acting on A_b, and the gauge parameter is hence dimensionless. The Eastwood–Singer functional (6.4.10), however, is obtained from the third derivatives of the potential, and hence leads to a sixth-order operator on A_b (see below). This is why, to preserve the correct dimensions, the gauge parameter can only be dimensionful. For convenience, we here write it as $\frac{\beta}{\alpha}$ as specified before, so that all numerical operations refer to a dimensionless

α. A double integration by parts is then necessary to express the Euclidean Lagrangian in the form $\frac{1}{2}A_b \mathcal{P}^{bc} A_c$, where

$$\mathcal{P}^{bc} \equiv -g^{bc}\Box + \left(1 - \frac{\beta}{\alpha}\Box^2\right)\nabla^b\nabla^c. \tag{6.4.11}$$

Note that β has dimension [length]^4. The operator $\mathcal{P}^{bc}$ is a complicated sixth-order elliptic operator, and it is unclear how to deal properly with it for finite values of α, since it leads to a complete departure from the standard scheme in quantum field theory, where the gauge-field operator receives contributions involving second-order derivatives from the gauge-invariant Lagrangian, and derivatives of order not greater than two from the gauge-averaging term, for bosonic field theories. Ultimately, its sixth-order nature might point out that the Eastwood–Singer gauge is incompatible with the attempt to perform averages over gauge functionals following the Faddeev–Popov method (Faddeev and Popov 1967, Esposito *et al.* 1997). However, in the limit as $\alpha \rightarrow \infty$, it 'tends' to the following second-order operator:

$$P^{bc} = -g^{bc}\Box + \nabla^b\nabla^c. \tag{6.4.12}$$

This operator is non-minimal, since the term $\nabla^b\nabla^c$ occurs, and has no Green's function, so that its leading symbol is degenerate. Thus, a consistent path-integral quantization (if admissible) is a hard task: one has to work, for finite values of α, with the sixth-order operator (6.4.11). The limit $\alpha \rightarrow \infty$ can only be taken *at the end of all calculations*, but not straight away. With this understanding, it is worth remarking that the operator (6.4.12) is also the limiting form, as $\alpha \rightarrow \infty$, of the gauge-field operator

$$-g^{bc}\Box + \left(1 - \frac{1}{\alpha}\right)\nabla^b\nabla^c,$$

obtained upon choosing the Lorenz gauge-averaging functional in flat space.

Of course, we need to specify boundary conditions on A_b and ghost perturbations. For this purpose, we put to zero at the boundary the whole set of A_b perturbations (Esposito 1997b):

$$\Big[A_b\Big]_{\partial M} = 0 \ \forall b = 0, 1, 2, 3, \tag{6.4.13}$$

and we require invariance of Eq. (6.4.13) under infinitesimal gauge transformations on A_b. This leads to (hereafter τ is a radial coordinate (Esposito *et al.* 1997))

$$[\varepsilon]_{\partial M} = 0, \tag{6.4.14}$$

$$\Big[\partial\varepsilon/\partial\tau\Big]_{\partial M} = 0. \tag{6.4.15}$$

Condition (6.4.14) results from the gauge invariance of the Dirichlet condition on A_k, $\forall k = 1,2,3$, and condition (6.4.15) results from the gauge invariance of the Dirichlet condition on A_0. Note that it would be inconsistent to impose the boundary conditions (6.4.13)–(6.4.15) when the Lorenz gauge-averaging functional is chosen, since the corresponding ghost operator is second-order.

When two boundary three-surfaces occur, Eqs. (6.4.14) and (6.4.15) lead to

$$[\varepsilon]_{\Sigma_1} = [\varepsilon]_{\Sigma_2} = 0, \tag{6.4.16}$$

$$\Big[\partial\varepsilon/\partial\tau\Big]_{\Sigma_1} = \Big[\partial\varepsilon/\partial\tau\Big]_{\Sigma_2} = 0. \tag{6.4.17}$$

When Eq. (6.4.10) is used, and the ghost operator is hence $\Box^2$, the four boundary conditions (6.4.16) and (6.4.17) provide enough conditions to determine completely the coefficients $C_1, ..., C_4$ in the linear combination

$$\varepsilon_{(\lambda)} = \sum_{i=1}^{4} C_i \rho_{i_{(\lambda)}}, \tag{6.4.18}$$

where $\rho_1, ..., \rho_4$ are the linearly independent solutions of the eigenvalue equation

$$\Box^2 \varepsilon_{(\lambda)} = \lambda\, \varepsilon_{(\lambda)}. \tag{6.4.19}$$

We therefore find that, when the conformally invariant gauge functionals (6.4.10) are used, the admissible boundary conditions differ substantially from the magnetic and electric schemes outlined in Eqs. (6.4.1)–(6.4.3) and (6.4.4)–(6.4.6), and are conformally invariant by construction.

Had we set to zero at the boundary A_k $(k = 1,2,3)$ and the functional (6.4.10), we would not have obtained enough boundary conditions for ghost perturbations, since both choices lead to Dirichlet conditions on the ghost. The boundary conditions (6.4.13) are also very important since they ensure the vanishing of all boundary terms resulting from integration by parts in the

Faddeev–Popov action. In the particular case when the three-surface Σ_1 shrinks to a point, which is relevant for (one-loop) quantum cosmology (Esposito *et al.* 1997), the boundary conditions read (here Σ is the bounding three-surface)

$$\Big[A_b\Big]_{\Sigma} = 0 \quad \forall b = 0, 1, 2, 3, \tag{6.4.20}$$

$$[\varepsilon]_{\Sigma} = 0, \tag{6.4.21}$$

$$\Big[\partial\varepsilon/\partial\tau\Big]_{\Sigma} = 0, \tag{6.4.22}$$

jointly with regularity at $\tau = 0$ of A_b, ε and $\frac{\partial\varepsilon}{\partial\tau}$. Many fascinating problems are now in sight. They are as follows.

(i) To prove uniqueness of the solution of the classical boundary-value problem

$$\Box^2 f = 0, \tag{6.4.23}$$

$$[f]_{\Sigma_1} = [f]_{\Sigma_2} = 0, \tag{6.4.24}$$

$$\Big[\partial f/\partial\tau\Big]_{\Sigma_1} = \Big[\partial f/\partial\tau\Big]_{\Sigma_2} = 0. \tag{6.4.25}$$

(ii) To study the quantum theory resulting from the operator (6.4.11) for finite values of α. Interestingly, no choice of α can then get rid of the sixth-order nature of the operator $\mathcal{P}^{bc}$. It is not clear, at least to the author, whether one should modify the gauge-averaging method in the Eastwood–Singer case. In the classical theory, if the Lorenz gauge condition is satisfied, the validity of the Eastwood–Singer gauge is automatically guaranteed, whereas the converse does not hold. Thus, the quantum theory is expected to reflect this property in a non-trivial way.

(iii) To evaluate the one-loop semiclassical approximation, at least in the presence of three-sphere boundaries. The ghost operator is then found to take the form

$$\Box^2 = \frac{\partial^4}{\partial\tau^4} + \frac{6}{\tau}\frac{\partial^3}{\partial\tau^3} + \frac{3}{\tau^2}\frac{\partial^2}{\partial\tau^2} - \frac{3}{\tau^3}\frac{\partial}{\partial\tau} + \frac{2}{\tau^2}\left(\frac{\partial^2}{\partial\tau^2} + \frac{1}{\tau}\frac{\partial}{\partial\tau}\right){}_{|i}{}^{|i} + \frac{1}{\tau^4}\left({}_{|i}{}^{|i}\right)^2. \tag{6.4.26}$$

With the notation of section 5.2, we denote by $|$ the operation of covariant differentiation tangentially with respect to the three-dimensional Levi–Civita connection of the boundary. If one expands the ghost perturbations on a family of three-spheres centred on the origin as (Esposito 1994a,b)

$$\varepsilon(x,\tau)=\sum_{n=1}^{\infty}\varepsilon_n(\tau)Q^{(n)}(x),$$

the operator (6.4.26), jointly with the properties of scalar harmonics, leads to the eigenvalue equation (cf. Eq. (6.4.19))

$$\frac{d^4\varepsilon_n}{d\tau^4}+\frac{6}{\tau}\frac{d^3\varepsilon_n}{d\tau^3}-\frac{(2n^2-5)}{\tau^2}\frac{d^2\varepsilon_n}{d\tau^2}$$
$$-\frac{(2n^2+1)}{\tau^3}\frac{d\varepsilon_n}{d\tau}+\left(\frac{(n^2-1)^2}{\tau^4}-\lambda_n\right)\varepsilon_n=0. \quad (6.4.27)$$

This equation admits a power-series solution in the form

$$\varepsilon_n(\tau)=\tau^{\rho}\sum_{k=0}^{\infty}b_{n,k}(n,k,\lambda_n)\tau^k. \quad (6.4.28)$$

The values of ρ are found by solving the fourth-order algebraic equation

$$\rho^4-2(n^2+1)\rho^2+(n^2-1)^2=0, \quad (6.4.29)$$

which has the four real roots $\pm(n\pm1)$. Moreover, the only non-vanishing $b_{n,k}$ coefficients are of the form $b_{n,4k}$, $\forall k=0,1,2,...$, and are given by (assuming that $b_{n,0}$ has been fixed)

$$b_{n,l}=\frac{\lambda_n\, b_{n,l-4}}{F(l,n,\rho)},\ \forall l=4,8,12,..., \quad (6.4.30)$$

where we have defined ($\forall k=0,1,2,...$)

$$F(k,n,\rho)\equiv(\rho+k)(\rho+k-1)(\rho+k-2)(\rho+k-3)$$
$$+6(\rho+k)(\rho+k-1)(\rho+k-2)$$
$$-(2n^2-5)(\rho+k)(\rho+k-1)$$
$$-(2n^2+1)(\rho+k)+(n^2-1)^2. \quad (6.4.31)$$

Since Eq. (6.4.27) is a fourth-order equation, it has four linearly independent integrals. One may expect that they can be obtained from Bessel functions, since we have 'squared up' the $\Box$ operator. Indeed, if one looks for solutions in the form (hereafter, $M\equiv\lambda_n^{\frac{1}{4}}$)

$$\varepsilon_n(\tau)=\frac{B_\nu(M\tau)}{\tau}, \quad (6.4.32)$$

where B_ν is a Bessel function of order ν depending on n, one finds that Eq. (6.4.27) becomes

$$\frac{M^4}{\tau}\left[\frac{d^4}{d\tau^4}+\frac{2}{M\tau}\frac{d^3}{d\tau^3}-\frac{(2n^2+1)}{M^2\tau^2}\frac{d^2}{d\tau^2}\right.$$
$$\left.+\frac{(2n^2+1)}{M^3\tau^3}\frac{d}{d\tau}+\left(\frac{n^4-4n^2}{M^4\tau^4}-1\right)\right]B_\nu=0. \qquad (6.4.33)$$

Remarkably, the operator in square brackets in Eq. (6.4.33) coincides with the operator occurring in the fourth-order differential equation obeyed by Bessel functions $B_\nu(M\tau)$ of order $\nu=n$ (see section 3.5 of Magnus *et al.* (1966)). One can thus write

$$\varepsilon_n(\tau)=C_{1,n}\frac{J_n(M\tau)}{\tau}+C_{2,n}\frac{N_n(M\tau)}{\tau}$$
$$+C_{3,n}\frac{I_n(M\tau)}{\tau}+C_{4,n}\frac{K_n(M\tau)}{\tau}. \qquad (6.4.34)$$

It remains to be seen, however, what is the relation between Eq. (6.4.34) and the solution in the form (6.4.28) when ρ equals the roots $\rho_+\equiv n+1$ and $\rho_-\equiv -n+1$, for which (cf. Eq. (6.4.31))

$$F(k,n,\rho_\pm)=(k\pm n)^4+4(k\pm n)^3-(2n^2-4)(k\pm n)^2$$
$$-4n^2(k\pm n)+n^4-4n^2. \qquad (6.4.35)$$

The eigenvalue condition is eventually obtained by imposing that the determinant of a 4×4 matrix should vanish. This is necessary to find non-trivial solutions for the coefficients $C_{1,n}, C_{2,n}, C_{3,n}$ and $C_{4,n}$ in (6.4.34), once that the boundary conditions (6.4.21) and (6.4.22) are imposed on concentric three-sphere boundaries.

(iv) To include the effects of curvature. As shown in Eastwood and Singer (1985), if the background four-geometry is curved, with Riemann tensor $R^a{}_{bcd}$, the conformally invariant gauge-averaging functional reads (cf. Eq. (6.4.10))

$$\Phi(A)\equiv\Box\nabla^b A_b+\nabla_c\left[\left(-2R^{bc}+\frac{2}{3}Rg^{bc}\right)A_b\right]. \qquad (6.4.36)$$

It would be interesting to study the (one-loop) quantum theory, at least when $\alpha\rightarrow\infty$, on curved backgrounds like S^4, which is relevant for inflation (Esposito *et al.* 1997), or $S^2\times S^2$, which is relevant for the bubbles picture in Euclidean quantum gravity, as proposed in Hawking (1996).

The form (6.4.11) of the differential operator on perturbations of the electromagnetic potential in the quantum theory, jointly with the boundary conditions (6.4.13)–(6.4.15), and the analytic solution (6.4.28)–(6.4.31) and (6.4.34) for ghost basis functions, are a slight extension of the work in Esposito (1997b). In the light of what we said so far, quantization via path integrals in conformally invariant gauges leads to severe technical problems, which are now under investigation for the first time. This, in turn, seems to add evidence in favour of Euclidean quantum gravity raising deep issues in quantum field theory (Esposito *et al.* 1997).

6.5 Conformally covariant operators

This section, motivated by the previous example of the conformal gauge for Maxwell theory, is devoted to a brief review of some key properties of conformally covariant operators. An important example is provided by the Paneitz operator. The framework for the introduction of this operator (Paneitz 1983) is a Riemannian manifold (M, g) of dimension $m \geq 3$ with Riemann curvature R, Ricci curvature ρ and scalar curvature τ. Denoting by d the exterior derivative, and by δ its formal adjoint, the Laplace operator on scalar functions reads

$$\Delta \equiv \delta \, d. \tag{6.5.1}$$

Moreover, one can consider (Branson 1996)

$$J \equiv \frac{\tau}{2(m-1)}, \tag{6.5.2}$$

$$V \equiv \frac{(\rho - Jg)}{(m-2)}, \tag{6.5.3}$$

$$T \equiv (m-2)J - 4V\cdot, \tag{6.5.4}$$

$$Q \equiv \frac{m}{2}J^2 - 2|\, V \,|^2 + \Delta J. \tag{6.5.5}$$

The operation $V\cdot$ is defined by

$$\Big(V \cdot \varphi\Big)_a \equiv V_a^{\; b} \, \varphi_b, \tag{6.5.6}$$

while $|\, V \,|^2 \equiv V^{cd} V_{cd}$, with the indices a, b, c, d ranging from 1 through m. The Weyl tensor, i.e. the part of Riemann which is invariant under conformal rescalings of the metric (and such that all its contractions vanish) can be therefore expressed as

$$C^a_{\ bcd} = R^a_{\ bcd} + V_{bc}\,\delta^a_{\ d} - V_{bd}\,\delta^a_{\ c} + V^a_{\ d}\,g_{bc} - V^a_{\ c}\,g_{bd}. \tag{6.5.7}$$

The Paneitz operator is, by definition, the following fourth-order operator:

$$P \equiv \Delta^2 + \delta T d + \frac{(m-4)}{2} Q. \tag{6.5.8}$$

This operator is *conformally covariant* (Erdmenger 1997) in that, if the metric is rescaled according to (cf. (5.3.1))

$$g_\omega \equiv e^{2\omega}\, g_0, \tag{6.5.9}$$

with $\omega \in C^\infty(M)$, then P rescales as

$$P_\omega = e^{-(m+4)\omega/2}\, P_0 \left[e^{(m-4)\omega/2}\right], \tag{6.5.10}$$

where square brackets are used to denote a multiplication operator, $[F]$, for any $F \in C^\infty(M)$. Indeed, there is an analogy with the construction of the *conformal Laplacian*, defined by

$$Y \equiv \Delta + \frac{(m-2)}{2} J, \tag{6.5.11}$$

which is conformally covariant in that

$$Y_\omega = e^{-(m+2)\omega/2}\, Y_0 \left[e^{(m-2)\omega/2}\right]. \tag{6.5.12}$$

If A is a formally self-adjoint differential operator with positive-definite leading symbol, its order can only be an even number, say $2l > 0$. For a smooth function $f \in C^\infty(M)$, the L^2 trace of fe^{-tA} has an asymptotic expansion as $t \to 0^+$ (cf. section 5.2)

$$\mathrm{Tr}_{L^2}\left(fe^{-tA}\right) \sim \sum_{k=0}^{\infty} t^{(k-m)/2l} \int_M fU_j[A]\, dv, \tag{6.5.13}$$

where dv is the Riemannian measure on M, and the $U_j[A]$ are invariants consisting of a polynomial in the total symbol, whose coefficients are smooth in the leading symbol (Branson 1996). Since M is here assumed to be a manifold without boundary, $U_k[A]$ vanishes for all odd values of k. The condition of conformal covariance for the operator A is that, under the conformal rescaling (6.5.9), A rescales according to the relation

$$\overline{A}_\omega = e^{-b\omega}\, A_0\, [e^{a\omega}], \tag{6.5.14}$$

for some real parameters a and b.

For the Paneitz operator, the invariant $U_4[P]$ is expressed, after introducing

$$u(m) \equiv 720(m^2-4)(4\pi)^{m/2}\frac{\Gamma((m-4)/2)}{\Gamma((m-4)/4)}, \tag{6.5.15}$$

by (Branson 1996)

$$U_4 = u^{-1}\Big\{-(m-8)(m+4)(m+2)(m-12)\Big[\Delta J + \frac{1}{2}(m-4)J^2\Big] + (m-8)\Big(m^3-52m-24\Big)Q + 2(m^2-4)\mid C\mid^2\Big\}. \tag{6.5.16}$$

In general, if A is conformally covariant, or a power of such (Eastwood and Rice 1987, Baston and Eastwood 1990, Graham *et al.* 1992, Branson 1995), the number of negative eigenvalues (counted, of course, with their multiplicity), and the multiplicity of 0 as an eigenvalue, are conformal invariants. One can then define the determinant of A by

$$-\log \mid \det A \mid \equiv \zeta_A'(0), \tag{6.5.17}$$

with

$$\text{sign} \det A = \text{number of } \{\lambda_k = 0\}, \tag{6.5.18}$$

where $\zeta_A(s)$ is the analytic continuation to the complex plane of

$$\sum_{\lambda_j \neq 0} \mid \lambda_j \mid^{-s}.$$

One can then study det A as a functional on the conformal class determined by the rescalings (6.5.9), by virtue of the relation

$$\zeta_{A_\omega}'(0) - \zeta_{A_0}'(0) = -\log \frac{\det A_\omega}{\det A_0}. \tag{6.5.19}$$

The problem of finding the extremals of the functional determinant within a conformal class is now under intensive investigation (Branson 1996).

Another important reference on conformal geometry and conformally covariant operators is the work by Parker and Rosenberg (1987). More recent developments are described in the work by Avramidi (1997a,c), where that author has studied the structure of diagonal singularities of Green functions of partial differential operators of even order acting on elements of $C^\infty(V,M)$, i.e. smooth sections of a vector bundle V over a Riemannian manifold M. In particular, the work in Avramidi (1997a,c) has investigated operators obtained by composition of second-order operators of Laplace type (cf. our section 6.4). Interestingly, singularities of the corresponding Green functions have been expressed in terms of heat-kernel coefficients.

6.6 Euclidean quantum gravity

The main physical motivations for studying the problem of a quantum theory of gravity (as far as the author can see) are as follows.

(i) Gravity is responsible for the large-scale structure of the universe we live in, and Einstein's theory of general relativity provides a good model of space-time physics, satisfying both mathematical elegance and agreement with observations (Hawking and Ellis 1973, Hawking 1979, Will 1993).

(ii) The formalism of quantum field theory, despite the lack of a rigorous mathematical setting, has led to many exciting developments in the quantum theory of the electromagnetic field and Yang–Mills fields, leading in turn to new results in mathematics: geometry of four-manifolds, topological quantum field theories, etc. (Donaldson and Kronheimer 1990, Freed and Uhlenbeck 1991, 1995).

(iii) The application of perturbative or non-perturbative methods to the quantization of Einstein's gravity, or of its supersymmetric version, has shown that formidable technical problems always emerge at some stage, possibly pointing out that all known ideas and methods lead to a mathematically inconsistent formulation of quantum gravity (DeWitt 1965, Hawking 1979, van Nieuwenhuizen 1981, Ashtekar 1988, 1991, Esposito 1994a, 1995, D'Eath 1996, Esposito *et al.* 1997, Immirzi 1997).

Nevertheless, decades of efforts by thousands of scientists from all over the world do not seem to have been useless. In particular, the path-integral approach remains an essential ingredient of any attempt to understand the properties of the quantized gravitational field. The crucial point is that quantum mechanics is a physical theory whose predictions are of statistical nature. When one tries to 'combine' it with general relativity, one may thus expect to obtain a formalism where statistical concepts such as the partition function (Gibbons and Hawking 1977) find a natural place. This is indeed the case for the Euclidean field theories. This property is possibly even more important, in our opinion, than the opportunity to obtain a space-time covariant approach to quantization, via the sum over suitable classes of (or all) Riemannian four-geometries with their topologies. Moreover, one knows that the effective action provides, in principle, a tool for studying quantum theory as a theory of small disturbances of the underlying classical theory, as well as many non-perturbative properties in field theory (Jona-Lasinio 1964, DeWitt 1965, 1981, Esposito *et al.* 1997 and references therein). The basic object of a space-time covariant formulation of quantum gravity may be viewed as being the path-integral representation of the $\langle \text{out} \mid \text{in} \rangle$ amplitude (Hawking 1979), which involves the consideration of ghost fields which reflect the gauge freedom of the classical theory (see Esposito *et al.* (1997) and the many references therein). In particular, what seems to emerge is that the consideration of the elliptic boundary-value problems of quantum gravity sheds new light on the one-loop semiclassical approximation, which is the 'bridge' in between the classical world and the as yet unknown (full) quantum theory (Gibbons and Hawking 1993, Esposito *et al.* 1997). We shall thus focus on this part of the quantum gravity problem, i.e. the boundary conditions on metric perturbations, when a Riemannian four-manifold (say M) with boundary is considered (this may be a portion of flat Euclidean four-space, or part of the de Sitter four-sphere, or a more general curved background). To begin, let us assume that spatial components of metric perturbations, say h_{ij}, are set to zero at the boundary:

$$\Big[h_{ij}\Big]_{\partial M} = 0. \tag{6.6.1}$$

Of course, this is suggested by what one would do in linearized theory at the classical level. A basic ingredient in our analysis is that Eq. (6.6.1) should be preserved under infinitesimal diffeomorphisms on metric perturbations. Their action on h_{ij} reads (Esposito *et al.* 1995b, 1997, Avramidi *et al.* 1996)

$$^{\varphi}h_{ij} = h_{ij} + \varphi_{(i|j)} + K_{ij}\varphi_0, \tag{6.6.2}$$

where $\varphi_b dx^b$ is the ghost one-form. It is thus clear that, if the extrinsic-curvature tensor of ∂M does not vanish, a necessary and sufficient condition for the preservation of the boundary conditions (6.6.1) under the transformations (6.6.2) is that the following boundary conditions should be imposed on the normal and tangential components of the ghost one-form:

$$\Big[\varphi_0\Big]_{\partial M} = 0, \tag{6.6.3}$$

$$\Big[\varphi_i\Big]_{\partial M} = 0. \tag{6.6.4}$$

At this stage, the remaining set of boundary conditions on metric perturbations, whose invariance under infinitesimal diffeomorphisms

$$^{\varphi}h_{ab} \equiv h_{ab} + \nabla_{(a}\,\varphi_{b)} \tag{6.6.5}$$

is *again* guaranteed (for consistency) by Eqs. (6.6.3) and (6.6.4), involves setting to zero at the boundary the gauge-averaging functional, say $\Phi_a(h)$, in the Faddeev–Popov action for Euclidean quantum gravity (Esposito *et al.* 1997). What happens is that, under the transformations (6.6.5), one finds

$$\Phi_a(h) - \Phi_a(^{\varphi}h) = \mathcal{F}_a^{\ b}\,\varphi_b, \tag{6.6.6}$$

where $\mathcal{F}_a^{\ b}$ is the ghost operator. If now the ghost one-form is expanded into a complete set of eigenfunctions of $\mathcal{F}_a^{\ b}$, say $\left\{f_b^{(\lambda)}\right\}$, which obey the eigenvalue equations

$$\mathcal{F}_a^{\ b}\,f_b^{(\lambda)} = \lambda\,f_a^{(\lambda)}, \tag{6.6.7}$$

the boundary conditions

$$\Big[\Phi_a(h)\Big]_{\partial M} = 0 \tag{6.6.8}$$

turn out to be invariant under (6.6.5) if and only if

$$\Big[f_a^{(\lambda)}\Big]_{\partial M} = 0 \ \ \forall a = 0, 1, 2, 3, \tag{6.6.9}$$

which implies in turn that (6.6.3) and (6.6.4) should hold, bearing in mind that

$$\varphi_a = \sum_{\lambda} C_\lambda\,f_a^{(\lambda)}, \tag{6.6.10}$$

where C_λ are some coefficients. In other words, Dirichlet boundary conditions for the ghost operator are *sufficient* to ensure gauge invariance of the boundary conditions (6.6.1) and (6.6.8) for the (gauge) operator acting on metric perturbations. A more elegant proof, relying on the use of the leading symbol of elliptic operators, can be found in section 4 of Avramidi and Esposito (1997).

In particular, if a covariant gauge-averaging functional of the de Donder type is used, i.e.

$$\Phi_a(h) \equiv \nabla^b \Big(h_{ab} - \frac{1}{2} g_{ab} g^{cd} h_{cd} \Big), \tag{6.6.11}$$

the boundary conditions (6.6.8) lead to normal and tangential derivatives of the normal components $n^a n^b h_{ab}$ and $n^b h_{ab}$, where n^b denotes the inward-pointing normal to the boundary of M.

In geometric language, one can say that metric perturbations h_{ab} are sections of the vector bundle V of symmetric rank-two tensors on M. On using the tensor field q_{ab} occurring in Eq. (5.2.10), one can define the projection operator (Avramidi *et al.* 1996, Moss and Silva 1997)

$$\Pi_{ab}^{\ \ cd} \equiv q^{c}_{\ (a} \, q^{d}_{\ b)}, \tag{6.6.12}$$

and the boundary operator corresponding to (6.6.1) and (6.6.8) in the de Donder case may be split as the direct sum (cf. (5.4.2))

$$\mathcal{B} = \mathcal{B}_1 \oplus \mathcal{B}_2, \tag{6.6.13}$$

where $\mathcal{B}_1$ is proportional to $\Pi_{ab}^{\ \ cd}$, and takes into account the boundary conditions (we are writing (6.6.1) in a covariant form)

$$\Big[\Pi_{ab}^{\ \ cd} \, h_{cd} \Big]_{\partial M} = 0, \tag{6.6.14}$$

while $\mathcal{B}_2$ reads (Avramidi *et al.* 1996, Avramidi and Esposito 1998a)

$$\mathcal{B}_2 \equiv \Big(1\!\!1 - \Pi \Big) \Big[H \nabla_N + \frac{1}{2} \Big(\Gamma^i \widehat{\nabla}_i + \widehat{\nabla}_i \Gamma^i \Big) + S \Big], \tag{6.6.15}$$

where H is the metric on the bundle V:

$$H^{ab\ cd} \equiv g^{a(c} \, g^{d)b} - \frac{1}{2} g^{ab} g^{cd}, \tag{6.6.16}$$

and (e^i being a basis for $T(\partial M)$, and γ^{ij} being the induced metric on ∂M)

$$\Gamma^{i\ \ cd}_{\ ab} \equiv -N_a N_b e^{i(c} \, N^{d)} + N_{(a} \, e^i_{\ b)} N^c N^d, \tag{6.6.17}$$

$$S_{ab}^{\ \ cd} \equiv -N_a N_b N^c N^d K + 2 N_{(a} \, e^i_{\ b)} e^{j(c} \, N^{d)} \Big[K_{ij} + \gamma_{ij} K \Big]. \tag{6.6.18}$$

The correct way to interpret $\widehat{\nabla}_i \Gamma^i$ in (6.6.15) is via the Leibniz rule

$$\widehat{\nabla}_i \Gamma^i \, h = \Big(\widehat{\nabla}_i \Gamma^i \Big) h + \Gamma^i \widehat{\nabla}_i h.$$

At a deeper level, S is an endomorphism of the vector bundle, say $\widetilde{V}$, over ∂M, and Γ^i are endomorphism-valued vector fields over ∂M.

Note that, if H is replaced by the identity matrix, the operator in square brackets in (6.6.15) reduces to the form first studied in McAvity and Osborn (1991b):

$$\widetilde{\mathcal{B}} \equiv \nabla_N + \frac{1}{2}\left(\Gamma^i \widehat{\nabla}_i + \widehat{\nabla}_i \Gamma^i\right) + S. \qquad (6.6.19)$$

For simplicity, we consider, for the time being, a problem where the boundary operator is not a direct sum but consists only of the right-hand side of (6.6.19), and we assume that the following three properties hold (Avramidi and Esposito 1998a).

(i) The matrices Γ^i commute with each other:

$$[\Gamma^i, \Gamma^j] = 0. \qquad (6.6.20)$$

(ii) The matrix $\Gamma^2 \equiv \gamma_{ij}\Gamma^i\Gamma^j$, which automatically commutes with Γ^i by virtue of (6.6.20), commutes also with the matrix S:

$$[\Gamma^2, S] = 0. \qquad (6.6.21)$$

(iii) The matrices Γ^i are covariantly constant with respect to the induced connection on the boundary: $\widehat{\nabla}_i \Gamma^j = 0$.

At this stage, it is already clear that tangential derivatives lead to a highly non-trivial 'invariance theory' (Gilkey 1995). More precisely, we know from section 5.2, which relies in turn on Weyl's theory of the invariants of the orthogonal group (Weyl 1946, Gilkey 1995), that the integrand in the general formulae for interior terms in heat-kernel asymptotics is a polynomial which can be found using only tensor products and contraction of tensor arguments. Thus, by virtue of (6.6.19), it is clear that the integrand in the boundary term in (5.2.19) should be supplemented by

$$f \, K_{ij} \, \Gamma^i \, \Gamma^j,$$

since this is the only local invariant which is linear in the extrinsic curvature and is built from contractions of K_{ij} with Γ^i. The resulting a_1 coefficient can be denoted, for clarity, by $a_1(f, P, \widetilde{\mathcal{B}})$, where P is an operator of Laplace type (see (5.2.1)).

Similarly, to obtain the form of $a_{3/2}(f, P, \widetilde{\mathcal{B}})$ one has to consider, further to the invariants occurring in (5.2.20), all possible contractions of the matrices Γ^i with geometric objects of the form (K being the second fundamental form of the boundary)

$$fK^2 \ , \ fKS \ , \ f\hat{\nabla}K \ , \ f\hat{\nabla}S \ , \ fR \ , \ f\Omega \ , \ f_{;N}K.$$

As shown in Avramidi and Esposito (1998a), this leads to eight new invariants in the general formula for $a_{3/2}(f, P, \widetilde{\mathcal{B}})$. Explicitly, the integrand for this heat-kernel coefficient is a linear combination of

$$f(K_{ij}\Gamma^i\Gamma^j)^2, f(K_{ij}\Gamma^i\Gamma^j)K, fK_{il}K^l{}_j\Gamma^i\Gamma^j, fK_{ij}\Gamma^i\Gamma^j S,$$

and

$$fR_{iNjN}\Gamma^i\Gamma^j, fR^l{}_{ilj}\Gamma^i\Gamma^j,$$

$$f\Omega_{iN}\Gamma^i, f_{;N}K_{ij}\Gamma^i\Gamma^j.$$

The number of such invariants is rapidly increasing for higher-order heat-kernel coefficients, and the same method shows that 41 new invariants contribute to $a_2(f, P, \widetilde{\mathcal{B}})$ (Avramidi and Esposito 1998a). Moreover, all the invariants contributing to interior terms are weighted by *universal functions*, rather than the universal constants of section 5.2. By this we mean functions which depend on the local coordinates on the boundary, whose dependence on Γ^i is realized by means of functions of $\Gamma^2 \equiv \Gamma_i\Gamma^i$. Hence they are not affected by conformal rescalings of the metric, and their contribution to conformal-variation formulae is purely algebraic. This point was not appreciated in McAvity and Osborn (1991b), but is investigated in detail in Avramidi and Esposito (1998a) and Dowker and Kirsten (1997).

Let $\sigma_{l/2}$ be the linear combination of local invariants occurring in the integrand for the heat-kernel coefficient $a_{l/2}$. It is thus clear that the knowledge of all local invariants in $\sigma_{l/2}$ plays a role in the form of $\sigma_{(l+1)/2}$, and one can express the properties mentioned above in the form

$$\sigma_1 = f\sum_{i=1}^{i_1}\mathcal{U}_i^{(1,1)}I_i^{(1)}, \tag{6.6.22}$$

$$\sigma_{3/2} = f\sum_{i=1}^{i_2}\mathcal{U}_i^{(3/2,3/2)}I_i^{(3/2)} + f_{;N}\sum_{i=1}^{i_1}\mathcal{U}_i^{(3/2,1)}I_i^{(1)}, \tag{6.6.23}$$

$$\sigma_2 = f \sum_{i=1}^{i_3} \mathcal{U}_i^{(2,2)} I_i^{(2)} + f_{;N} \sum_{i=1}^{i_2} \mathcal{U}_i^{(2,3/2)} I_i^{(3/2)} + f_{;NN} \sum_{i=1}^{i_1} \mathcal{U}_i^{(2,1)} I_i^{(1)}. \tag{6.6.24}$$

With our notation, $i_1 = 1, i_2 = 7, i_3 = 33$, and $\mathcal{U}_i^{(x,y)}$ are the universal functions, where i is an integer ≥ 1, x is always equal to the order $l/2$ of $\sigma_{l/2}$, and y is equal to the label of the invariant $I_i^{(y)}$, which does not contain f or derivatives of f and is of dimension $2y-1$ in the extrinsic-curvature tensor of the boundary, or in tensors of the same dimension of the extrinsic curvature.

These remarks make it possible to write down a formula which holds for all $l \geq 2$:

$$\sigma_{l/2}(f, R, \Omega, K, E, \Gamma, S)$$
$$= \sum_{r=0}^{l-2} f^{(r)} \sum_{i=1}^{i_{l-r-1}} \mathcal{U}_i^{(l/2,(l-r)/2)}[\Gamma^2]$$
$$\cdot I_i^{(l-r)/2}[R, \Omega, K, E, \Gamma, S], \tag{6.6.25}$$

where $f^{(r)}$ is the normal derivative of f of order r (with $f^{(0)} = f$), and square brackets are used for the arguments of universal functions and local invariants, respectively. This parametrization of heat-kernel coefficients holds *provided that the three assumptions* (i)–(iii) *are satisfied.*

Many important problems remain unsolved. They are as follows.

(i) The conformal-variation method is, by itself, unable to determine all universal functions. Since tangential derivatives occur in the boundary operator (6.6.19), the universal functions obey an involved set of equations. What is lacking at present is a systematic algorithm for the generation of all such equations. This achievement, jointly with the conformal-variation formulae (see appendix 6.C), should lead in turn to a complete understanding of heat-kernel asymptotics with the generalized boundary operator (6.6.19) in the Abelian case. At present, in such a case, only the a_1 coefficient is completely determined (McAvity and Osborn 1991b, Avramidi and Esposito 1998a, Dowker and Kirsten 1997). In the evaluation of $a_{3/2}$, 18 unknown universal functions occur, but so far only 11 equations among them have been derived (Avramidi and Esposito 1998a). Nevertheless, this is a non-trivial intermediate step, and there is hope to determine completely the form of $a_{3/2}$ in this particular case.

(ii) Euclidean quantum gravity, however, needs much more, since, with Barvinsky boundary conditions (Barvinsky 1987, Esposito *et al.* 1995b, 1997, Avramidi *et al.* 1996), the boundary operator is expressed by the direct sum (6.6.13), with $\mathcal{B}_2$ given by Eq. (6.6.15) in the de Donder gauge. As shown in Avramidi and Esposito (1998a), the condition (6.6.20) no longer holds in this case. Moreover, recent work by Avramidi and Esposito (1998b) has proved that the strong ellipticity condition described in appendix 3.A fails to hold when Barvinsky boundary conditions are imposed and the de Donder gauge-averaging functional is used. This is a highly undesirable property, because it means that the spectrum of the graviton operator becomes infinitely degenerate in the neighbourhood of the zero-eigenvalue. Such a result leads in turn, in our opinion, to a deep crisis in the current attempts to understand Euclidean quantum gravity on manifolds with boundary (Avramidi and Esposito 1997, 1998b). More precisely, the analysis in Avramidi and Esposito (1997, 1998b) proves that it is impossible to satisfy all the following requirements.

(1) An operator on metric perturbations, say P, of Laplace type.

(2) Local nature of the boundary operator B_P.

(3) Gauge invariance of the boundary conditions.

(4) Strong ellipticity of the boundary-value problem (P, B_P).

When the first three conditions hold, the fourth is not satisfied. In physical applications, the strong ellipticity of the boundary-value problem is crucial to ensure the existence of the asymptotic expansions frequently studied in the theory of heat kernels. When the condition is violated, the semiclassical expansion of the effective action is no longer well defined. The explicit calculations in section 7 of Avramidi and Esposito (1997) show that, when strong ellipticity is not respected, the heat-kernel diagonal in Euclidean quantum gravity acquires a part which is not integrable on the boundary of the manifold. Among the possible ways out, one might consider the possibility that, on taking gauge-averaging functionals that lead to non-minimal operators on metric perturbations, strong ellipticity is recovered also for the gravitational field with gauge-invariant boundary conditions. However, the corresponding analysis is even more complicated than the one performed in Avramidi and Esposito (1997, 1998b), and there is no *a priori* argument showing why the consideration of non-minimal operators might be more helpful. So far, strong ellipticity for non-minimal operators has been proved only in simpler boundary-value problems by Gilkey (1995). A further task for theoretical and mathematical physicists is to understand the deeper underlying reasons (if any) for the operator on metric perturbations being symmetric with gauge-invariant boundary conditions, as proved in detail in Avramidi *et al.* (1996), but still failing to satisfy the strong ellipticity condition, as found in Avramidi and Esposito (1997, 1998b).

(iii) Barvinsky boundary conditions turn out to be a particular case of the most general set of BRST-invariant boundary conditions, derived recently by Moss and Silva (1997). The heat-kernel asymptotics corresponding to Moss–Silva boundary conditions remains unknown. For example, when the commutation property (6.6.20) is not respected, only the coefficient $a_{1/2}$ is known, following section 3 of Avramidi and Esposito (1997).

(iv) At the mathematical level, there is also the problem of heat-content asymptotics when the boundary operator takes the form (6.6.19). Within this framework, for a given smooth vector bundle V over M, with dual bundle V^*, denoting by $\langle \cdot, \cdot \rangle$ the natural pairing between V and V^*, with $f_1 \in C^\infty(V)$ and $f_2 \in C^\infty(V^*)$, one defines (see (5.2.1))

$$\gamma(f_1, f_2, P, B)(t) \equiv \langle e^{-tP} f_1, f_2 \rangle_{L^2}, \tag{6.6.26}$$

where B is the boundary operator. Following van den Berg and Gilkey (1994) and Gilkey (1995), one knows that, as $t \to 0^+$, γ has an asymptotic expansion

$$\gamma(f_1, f_2, P, B) \sim \sum_{k=0}^{\infty} t^{k/2} \, \gamma_k(f_1, f_2, P, B). \tag{6.6.27}$$

For Dirichlet or Neumann boundary conditions, there exist local invariants such that

$$\gamma_k(f_1, f_2, P, B) = \gamma_k^{(int)}(f_1, f_2, P)[M] + \gamma_k^{(bd)}(f_1, f_2, P, B)[\partial M]. \tag{6.6.28}$$

This corresponds to the (physical) problem of finding the asymptotic expansion of the total heat content of M for a given initial temperature. However, the invariants contributing to Eq. (6.6.28) when the boundary operator B coincides with (6.6.19) or with the Moss–Silva boundary operators, have not yet been studied.

From the point of view of theoretical physics, these investigations remain very valuable, since they show that, upon viewing quantum field theory as a theory of small disturbances of the underlying classical theory (DeWitt 1965, 1981, Esposito *et al.* 1997), the problem of finding the first non-trivial corrections to classical field theory is a problem in invariance theory, at least when (6.6.20) and (6.6.21) hold, and the result may be expressed in purely geometric terms. Whether this will be enough to begin a (new) series of substantial achievements in quantum gravity, for which the technical problems, as we stressed, are formidable, is a fascinating problem for the years to come.

6.7 Dirac operator and the theory of four-manifolds

Over the past few years, the formalism of spinors and of the Dirac operator has been applied to get new insight into quantum field theory on the one hand, and the mathematical theory of four-manifolds on the other hand. In the original paper by Seiberg and Witten (1994), the authors studied four-dimensional $N = 2$ supersymmetric gauge theories with matter multiplets. For all such models with gauge group given by $SU(2)$, they derived the exact metric on the moduli space of quantum vacua, and the exact spectrum of stable massive states. New physical phenomena were found to occur, in particular, a chiral symmetry breaking driven by the condensation of magnetic monopoles. Interestingly, whenever conformal invariance is broken only by mass terms, their formalism leads to results which are invariant under electric-magnetic duality.

The mathematical analysis of the structures used by Seiberg and Witten and of some of their results is, by itself, quite enlightening. Relying on the presentation by Donaldson (1996), some key elements can be described as follows. Suppose that M is an oriented Riemannian four-manifold. If a spin-structure on M exists (see section 1.3), one deals with a pair of complex bundles S^+ and S^- over M, each with structure group $SU(2)$ and related to the tangent bundle $T(M)$ by a structure map

$$c : T(M) \rightarrow \mathrm{Hom}(S^+, S^-).$$

The map c is the symbol of the partial Dirac operator (see the discussion leading to (1.4.13))

$$D : \Gamma(S^+) \rightarrow \Gamma(S^-),$$

and the Lichnerowicz–Weitzenböck formula for the squared Dirac operator states that

$$D^* D \psi = \nabla^* \nabla \psi + \frac{R}{4} \psi. \tag{6.7.1}$$

With a standard notation, ∇ is the covariant derivative on spinors, induced by the Levi–Civita connection (see (1.3.2)), and R is the scalar curvature, whose action on ψ is multiplicative. One might now consider an auxiliary bundle over M, say E, with a Hermitian metric. The corresponding spinors are sections of the tensor-product bundles $S^\pm \otimes E$ (see (3.3.6)), and the (squared) Dirac operator on these coupled spinors satisfies the equation

$$D^* D \psi = \nabla^* \nabla \psi + \frac{R}{4} \psi - F_E^+(\psi), \tag{6.7.2}$$

where

$$F_E^+ \equiv \frac{1}{2}(F_E + *F_E) \tag{6.7.3}$$

is the self-dual part of the curvature of E.

As we know from chapter 1, spin-structures on M may not exist globally, the obstruction being given by the second Stiefel–Whitney class $w_2(M)$. However, even when spin-structures on M are not available, one can build the so-called $Spin^c$ structures. A $Spin^c$ structure consists of a pair of vector bundles over M, say W^+ and W^-, jointly with an isomorphism

$$\Lambda^2 W^+ \cong \Lambda^2 W^- \equiv L, \tag{6.7.4}$$

such that, *locally*,

$$W^\pm = S^\pm \otimes \mathcal{U}, \tag{6.7.5}$$

where $\mathcal{U}$ is a 'local square root' of L, in that

$$\mathcal{U} \otimes \mathcal{U} = L. \tag{6.7.6}$$

By virtue of the work of Hirzebruch and Hopf, one knows that $Spin^c$ structures exist on *any oriented four-manifold.* For a given connection on L, one has the Dirac operator

$$D : \Gamma(W^+) \rightarrow \Gamma(W^-),$$

which coincides, locally, with the Dirac operator on $\mathcal{U}$-valued spinors. For this Dirac operator, the Lichnerowicz formula (6.7.2) reads

$$D^* D\psi = \nabla^* \nabla \psi + \frac{R}{4}\psi - \frac{1}{2} F_L^+(\psi), \tag{6.7.7}$$

where the factor $\frac{1}{2}$ results from taking the square root of L, in the sense specified by (6.7.6).

The *Seiberg–Witten equations* for a four-manifold M with $Spin^c$ structure $W^\pm$ are equations for a pair (A, ψ), where

(i) A is a unitary connection on $L = \Lambda^2 W^\pm$,

(ii) ψ is a section of W^+.

For a pair of elements of W^+, say ξ and η, one writes $\xi\eta^*$ for the endomorphism $\theta \rightarrow \langle\theta, \eta\rangle\xi$ of W^+. This endomorphism has a trace-free part which lies in the image of the map ρ from the self-dual two-forms Λ^+ to $\text{Hom}(S^+, S^-)$. The corresponding element of $\Lambda^+ \otimes C$ is denoted by $\tau(\xi, \eta)$. In other words, τ is the sesquilinear map

$$\tau : W^+ \times W^+ \rightarrow \Lambda^+ \otimes C.$$

After the specification of all these geometric data, the Seiberg–Witten equations can be written as the system (Witten 1994)

$$D_A \psi = 0, \tag{6.7.8}$$

$$F_A^+ = -\tau(\psi, \psi). \tag{6.7.9}$$

The sign on the right-hand of Eq. (6.7.9) is really crucial. To understand thoroughly this property, consider a solution (A, ψ) of the Seiberg–Witten equations over a *compact* four-manifold M. One then finds, from Eqs. (6.7.8) and (6.7.9), that

$$0 = D_A^* D_A \psi = \nabla_A^* \nabla_A \psi - \frac{1}{2} F_A^+(\psi) + \frac{R}{4}\psi. \tag{6.7.10}$$

On taking the L^2 inner product with ψ, one obtains

$$\int_M \left[| \nabla_A \psi |^2 - \frac{1}{2}(F_A^+(\psi), \psi) + \frac{R}{4} | \psi |^2 \right] d\mu = 0. \tag{6.7.11}$$

Moreover, by virtue of Eq. (6.7.9), one has

$$(F_A^+(\psi), \psi) = -(\tau(\psi, \psi)(\psi), \psi), \tag{6.7.12}$$

where, by construction,

$$\tau(\psi, \psi)(\psi) = \frac{1}{2} | \psi |^2 \psi. \tag{6.7.13}$$

The equality (6.7.13) is easily proved if one thinks of the spinor ψ as a column vector $\begin{pmatrix} a \\ b \end{pmatrix}$, because this yields

$$| \psi |^2 = a\overline{a} + b\overline{b} = | a |^2 + | b |^2, \tag{6.7.14}$$

$$\tau(\psi, \psi)(\psi) = \begin{pmatrix} \frac{1}{2}(| a |^2 - | b |^2) & a\overline{b} \\ b\overline{a} & \frac{1}{2}(| b |^2 - | a |^2) \end{pmatrix} \begin{pmatrix} a \\ b \end{pmatrix}, \tag{6.7.15}$$

and (6.7.15) coincides with the right-hand side of (6.7.13) by elementary algebra. Now by virtue of (6.7.12) and (6.7.13), the equation (6.7.11) is re-expressed in the form

$$\int_M \left[| \nabla_A \psi |^2 + \frac{1}{4} | \psi |^4 \right] d\mu = - \int_M \frac{R}{4} | \psi |^2 d\mu. \tag{6.7.16}$$

In particular, if the scalar curvature of M is everywhere non-negative, the only solution of the Seiberg–Witten equation occurs when

$$\psi = 0, \tag{6.7.17}$$

$$F_A^+ = 0, \tag{6.7.18}$$

which are also the solutions of the $U(1)$ instanton equation.

If no assumption is made on the sign of the scalar curvature, one can still obtain some information from Eq. (6.7.16). For this purpose, denoting by $-c \leq 0$ the minimum of the scalar curvature R over M, one finds

$$\int_M \left[| \nabla_A \psi |^2 + \frac{1}{4} | \psi |^4 \right] d\mu \leq \frac{c}{4} \int_M | \psi |^2 d\mu$$

$$\leq \frac{c}{4} (\mathrm{Vol}(M))^{1/2} \sqrt{\int_M | \psi |^4 d\mu}, \tag{6.7.19}$$

by virtue of the Cauchy–Schwarz inequality. In the inequality

(6.7.19), the integral of $|\psi|^4$ is bounded by $c^2 \text{Vol}(M)$, and hence one gets eventually a bound on the L^2 norm of F_A^+ in terms of the geometry of M.

A variational derivation of the Seiberg–Witten equations can be obtained. Indeed, for any pair (A, ψ), one may consider the integral

$$E(A, \psi) \equiv \int_M \left[|\nabla_A \psi|^2 + \frac{1}{2} |F_A|^2 + \frac{1}{8} (|\psi|^2 + R)^2 \right] d\mu. \tag{6.7.20}$$

One can now use Eq. (6.7.7) to express

$$\nabla_A^* \nabla_A \psi = D_A^* D_A \psi + \frac{1}{2} F_A^+(\psi) - \frac{R}{4} \psi. \tag{6.7.21}$$

Moreover, there exists a topological invariant $C_1^2(L) \in H^4(M; Z)$ which has the integral representation

$$C_1^2(L) = \frac{1}{4\pi^2} \int_M \left[|F_A^+|^2 - |F_A^-|^2 \right] d\mu. \tag{6.7.22}$$

Thus, if in (6.7.21) one takes the L^2 inner product with ψ, the insertion into (6.7.20) leads to a cancellation of the $R|\psi|^2$ term in the integrand for the functional $E(A, \psi)$, which reads

$$E(A, \psi) = \int_M \left[|D_A \psi|^2 + |F_A^+ + \tau(\psi, \psi)|^2 \right] d\mu + E_0(M, L), \tag{6.7.23}$$

where

$$E_0(M, L) \equiv \int_M \frac{R^2}{8} d\mu + 2\pi^2 C_1^2(L). \tag{6.7.24}$$

The Seiberg–Witten equations (6.7.8) and (6.7.9) are hence found to provide the absolute minima of the functional $E(A, \psi)$. In the two-component spinor notation used in the second part of chapter 4, they can be cast in the form (due to LeBrun)

$$\left(\nabla_{AA'} + \frac{1}{2} \theta_{AA'} \right) \Phi^A = 0, \tag{6.7.25}$$

$$\nabla^{A'}_{\ \ (A} \, \theta_{B)A'} = \frac{1}{\sqrt{2}} \Phi_{(A} \, \overline{\Phi}_{B)}. \tag{6.7.26}$$

Remarkably, the Seiberg–Witten equations define differential-topological invariants of the underlying four-manifold M, in that the following properties hold (Donaldson 1996).

(i) The linearization of the equations about a solution is represented by linear elliptic differential equations. This is the Fredholm property, and the index of the linearized equation provides the dimension of the solutions space.

(ii) The space of solutions is compact.

(iii) No reducible solutions can be found in generic one-parameter families of equations.

(iv) Orientability can be established (see below).

To check property (i), the quadratic term $\tau(\psi, \psi)$ in Eq. (6.7.9) can be ignored, because it does not affect the leading symbol, say σ_L, of the linearization. At the level of σ_L, the linearization consists of the linearization of the $U(1)$ instanton equation, represented by the operator $(d^* + d^+)$ acting on ordinary forms, jointly with the Dirac operator D_A. Each of these operators is elliptic, and hence the index of the problem is expressed by invariants in the form

$$\begin{aligned} i(L) &= \mathrm{ind}(d^* \oplus d^+) + \mathrm{ind}(D_A) \\ &= -\frac{1}{2}(\chi(M) + \tau(M)) + \frac{1}{4}\Big(C_1^2(L) - \tau(M)\Big) \\ &= \frac{1}{4}\Big(C_1^2(L) - 2\chi(M) - 3\tau(M)\Big), \end{aligned} \tag{6.7.27}$$

where χ and τ are the Euler characteristic and the signature, respectively, and the index of D_A has been obtained with the help of the Atiyah–Singer index theorem. The number $i(L)$ represents the dimension of the Seiberg–Witten solution space. A class of perturbations of the Seiberg–Witten equations are easily obtained, on replacing Eq. (6.7.9) by the equation

$$F_A^+ = -\tau(\psi, \psi) + \theta, \tag{6.7.28}$$

where θ is *any* self-dual two-form on M. One can then find arbitrarily small θ such that the space of solutions of the perturbed equation is a manifold of dimension $i(L)$ (Witten 1994).

As far as compactness is concerned, this follows from *a priori* estimates on the solutions, which can be obtained, in turn, from energy estimates, by applying the maximum principle to second-order equations. Indeed, if (A, ψ) is a Seiberg–Witten solution, then it follows from Eq. (6.7.10) that

$$\nabla_A^* \nabla_A \psi = \frac{1}{2} F_A^+(\psi) - \frac{R}{4}\psi. \qquad (6.7.29)$$

At this stage, bearing in mind the Eqs. (6.7.12) and (6.7.13), one finds the inequality

$$2 \triangle (|\psi|^2) \leq c|\psi|^2 - |\psi|^4, \qquad (6.7.30)$$

where $-c$ is the lower bound on the scalar curvature. When $|\psi|$ is maximal, one has $\triangle(|\psi|^2) \geq 0$, which then implies, to be consistent with (6.7.30), that ψ obeys the restriction

$$\max|\psi| \leq \sqrt{c}. \qquad (6.7.31)$$

On inserting this inequality into the equations, one gets an L^∞ bound on F_A^+, and elliptic theory can be used to obtain an L^p bound on the full curvature $F(A)$, for all values of p. This property leads, in turn, to compactness of the space of solutions.

Note now that, if a non-trivial gauge transformation, say h (which is an element of $\mathrm{Aut}(L)$) fixes a pair (A, ψ), then ψ has to vanish, and $h \in U(1)$ is forced to be a constant scalar. Thus, the only reducible Seiberg–Witten solutions are the self-dual $U(1)$ connections. On the other hand, these do not occur in generic r-dimensional families of metrics on M, provided that $b^+(M) > r$ (b^+ being the dimension of the space of self-dual harmonic forms). Thus, in particular, if $b^+ > 1$, statement (iii) is correct.

By orientation one means orientation of the moduli space (cf. appendix 6.D), and this is provided by a 'line' of a suitable bundle over the space $\mathcal{C}^*$ of all irreducible pairs (A, ψ) modulo gauge transformations. This line splits into the tensor product of a contribution resulting from the $(d^* + d^+)$ operator on ordinary forms, and a contribution obtained from the Dirac operator. The former inherits the orientation from det $\mathrm{ind}(d^* + d^+)$, and the latter acquires an orientation because the spinors are described in terms of complex vector bundles, and the Dirac operator has a complex-linear action (see section 1.4). Interestingly, the orientation of the Seiberg–Witten moduli space requires therefore the same data as for the instanton spaces (Donaldson 1996).

Among the many mathematical applications of the Seiberg–Witten equations, which are described in detail in Donaldson (1996), we would like to mention the ability to distinguish differentiable four-manifolds within the same homeomorphism type.

For example, one can prove that connected sums of p copies of the complex projective space CP^2 and q copies of $\overline{CP^2}$, with $q > 1$, and for which the Seiberg–Witten invariants vanish, are not diffeomorphic to Kähler surfaces. A crucial property is that *no four-manifold with $b^+ > 1$ and non-zero Seiberg–Witten invariants admits a metric of positive scalar curvature.* By contrast, the connected sums mentioned before are known to admit such metrics. Hence it is clear that the classification of four-manifolds with metrics of positive scalar curvature is substantially different from the higher-dimensional case (where the only obstructions result from cobordism theory and characteristic classes). Moreover, it has been proved that no three-manifold with positive scalar curvature can be imbedded in a four-manifold with non-zero Seiberg–Witten invariants, in such a way that $b^+ > 0$ on each side. Another important result is due to LeBrun (1995). He has proved that *any Einstein four-manifold with non-zero Seiberg–Witten invariants* satisfies the inequality

$$\chi - 3\tau \geq 0. \tag{6.7.32}$$

With the exception of complex surfaces, this is a new restriction on Einstein metrics.

It remains to be seen, however, whether the large amount of new mathematical developments in this area of research (Morgan 1996) will have some impact on the longstanding problems of quantum gravity (cf. section 6.6), and to what extent the final picture (if any) on four-dimensional geometry and topology will incorporate the present results.

Appendix 6.A

The structure of the $\zeta(0)$ calculation resulting from the eigenvalue condition (6.3.4) is so interesting that we find it appropriate, for completeness, to present a brief outline in this appendix, although a more extensive treatment can be found in D'Eath and Esposito (1991a), Esposito (1994a), Esposito *et al.* (1997).

The function F occurring in Eq. (6.3.4) can be expressed in terms of its zeroes μ_i as (ρ being a constant)

$$F(z) = \rho \, z^{2(n+1)} \prod_{i=1}^{\infty} \left(1 - \frac{z^2}{\mu_i^2}\right). \tag{6.A.1}$$

Thus, setting $m \equiv n+2$, one finds

$$J_{m-1}^2(x) - J_m^2(x) = {J'_m}^2 + \left(\frac{m^2}{x^2} - 1\right) J_m^2 + 2\frac{m}{x} J_m J'_m. \quad (6.A.2)$$

Thus, on making the analytic continuation $x \rightarrow ix$ and then defining $\alpha_m \equiv \sqrt{m^2 + x^2}$, one obtains

$$\log\left[(ix)^{-2(m-1)}\left(J_{m-1}^2 - J_m^2\right)(ix)\right] \sim -\log(2\pi) + \log(\alpha_m) + 2\alpha_m - 2m\log(m+\alpha_m) + \log(\widetilde{\Sigma}). \quad (6.A.3)$$

In the asymptotic expansion (6.A.3), $\log(\widetilde{\Sigma})$ admits an asymptotic series in the form

$$\log(\widetilde{\Sigma}) \sim \left[\log(c_0) + \frac{A_1}{\alpha_m} + \frac{A_2}{\alpha_m^2} + \frac{A_3}{\alpha_m^3} + \ldots\right], \quad (6.A.4)$$

where, on using the Debye polynomials for uniform asymptotic expansions of Bessel functions (appendix 4.B), one finds (hereafter, $t \equiv \frac{m}{\alpha_m}$)

$$c_0 = 2(1+t), \quad (6.A.5)$$

$$A_1 = \sum_{r=0}^{2} k_{1r} t^r, \; A_2 = \sum_{r=0}^{4} k_{2r} t^r, \; A_3 = \sum_{r=0}^{6} k_{3r} t^r, \quad (6.A.6)$$

where

$$k_{10} = -\frac{1}{4}, \; k_{11} = 0, \; k_{12} = \frac{1}{12}, \quad (6.A.7)$$

$$k_{20} = 0, \; k_{21} = -\frac{1}{8}, \; k_{22} = k_{23} = \frac{1}{8}, \; k_{24} = -\frac{1}{8}, \quad (6.A.8)$$

$$k_{30} = \frac{5}{192}, \; k_{31} = -\frac{1}{8}, \; k_{32} = \frac{9}{320}, \; k_{33} = \frac{1}{2}, \quad (6.A.9)$$

$$k_{34} = -\frac{23}{64}, \; k_{35} = -\frac{3}{8}, \; k_{36} = \frac{179}{576}. \quad (6.A.10)$$

The corresponding ζ-function at large x (Moss 1989) has a uniform asymptotic expansion given by

$$\Gamma(3)\zeta(3, x^2) \sim W_\infty + \sum_{n=5}^{\infty} \hat{q}_n x^{-2-n}, \quad (6.A.11)$$

where, defining

$$S_1(m, \alpha_m(x)) \equiv -\log(\pi) + 2\alpha_m, \quad (6.A.12)$$

$$S_2(m, \alpha_m(x)) \equiv -(2m-1)\log(m+\alpha_m), \tag{6.A.13}$$

$$S_3(m, \alpha_m(x)) \equiv \sum_{r=0}^{2} k_{1r}\, m^r \alpha_m^{-r-1}, \tag{6.A.14}$$

$$S_4(m, \alpha_m(x)) \equiv \sum_{r=0}^{4} k_{2r}\, m^r \alpha_m^{-r-2}, \tag{6.A.15}$$

$$S_5(m, \alpha_m(x)) \equiv \sum_{r=0}^{6} k_{3r}\, m^r \alpha_m^{-r-3}, \tag{6.A.16}$$

W_∞ can be obtained as

$$W_\infty = \sum_{m=0}^{\infty}\left(m^2 - m\right)\left(\frac{1}{2x}\frac{d}{dx}\right)^3 \left[\sum_{i=1}^{5} S_i(m, \alpha_m(x))\right]. \tag{6.A.17}$$

The resulting $\zeta(0)$ value receives contributions from S_2, S_4 and S_5 only, and is given by

$$\zeta(0) = -\frac{1}{120} + \frac{1}{24} + \frac{1}{2}\sum_{r=0}^{4} k_{2r} - \frac{1}{2}\sum_{r=0}^{6} k_{3r} = \frac{11}{360}. \tag{6.A.18}$$

Of course, for a massless Dirac field, the full $\zeta(0)$ is twice the value in (6.A.18):

$$\zeta_{\text{Dirac}}(0) = \frac{11}{180}. \tag{6.A.19}$$

The same $\zeta(0)$ values are found for a massless spin-1/2 field in a four-sphere background bounded by a three-sphere. The detailed derivation of this property can be found in Kamenshchik and Mishakov (1993) (jointly with the analysis in the massive case). However, the agreement might have been expected on general ground, without performing any explicit calculation. The reason is, that for a conformally invariant problem the $\zeta(0)$ value is a conformal invariant (e.g. Branson and Gilkey 1994).

Appendix 6.B

Since we mention, in section 6.4, the Faddeev–Popov method of Gaussian averages over gauge functionals, we find it appropriate, for completeness, to emphasize some key points of this construction, while the reader is referred to DeWitt (1984), Esposito *et al.*

(1997), Avramidi and Esposito (1997) for an extensive description of the DeWitt formalism used hereafter.

For a given gauge theory with action functional S and generators of gauge transformations $\mathcal{R}_\alpha$, one has

$$\left[\mathcal{R}_\alpha, \mathcal{R}_\beta\right] S = \mathcal{R}_\alpha \mathcal{R}_\beta S - \mathcal{R}_\beta \mathcal{R}_\alpha S = 0. \tag{6.B.1}$$

The theory considers the manifold $\mathcal{M}$ of all fields, say φ. If, for all $\varphi \in \mathcal{M}$, the generators $\mathcal{R}_\alpha$ are linearly independent, and are linear in the fields, and form a complete set, then the most general solution of Eq. (6.B.1) reads

$$\left[\mathcal{R}_\alpha, \mathcal{R}_\beta\right] = C^\gamma_{\ \alpha\beta}\, \mathcal{R}_\gamma + T^{ij}_{\ \ \alpha\beta} \frac{\delta}{\delta\varphi^i} S_{,j}. \tag{6.B.2}$$

Hereafter, we assume that the $T^{ij}_{\ \ \alpha\beta}$ vanish $\forall i, j, \alpha, \beta$, and that the $C^\gamma_{\ \alpha\beta}$ are constant, in that they are annihilated by the operation of functional differentiation with respect to the fields. One then obtains a Lie-algebra structure for the commutator $\left[\mathcal{R}_\alpha, \mathcal{R}_\beta\right]$. Since we are studying a gauge theory, what is crucial is the existence of equivalence classes under the action of gauge transformations. The information on such equivalence classes is encoded into suitable 'coordinates', say (I^A, χ^α), obtained as follows. The I^A are *non-local functionals* which pick out the orbit where the field φ lies, whereas the $\chi^\alpha(\varphi)$ pick out the particular point on the orbit corresponding to the field φ. They are precisely the gauge functionals considered by Faddeev and Popov (1967). The physical $\langle \text{out} \mid \text{in} \rangle$ amplitude can be expressed, formally, by functional integration over equivalence classes (under gauge transformations) of field configurations:

$$\langle \text{out} \mid \text{in} \rangle = \int \mu(I) e^{iS[\varphi(I)]} dI. \tag{6.B.3}$$

It is now possible, at least formally, to re-express this abstract path-integral formula in terms of the original field variables. For this purpose, the integration over equivalence classes is made explicit by introducing the χ-integration with the help of a δ distribution, say $\delta(\chi(\varphi) - \zeta)$. After obtaining the Jacobian $J(\varphi)$ of the coordinate transformation from (I^A, χ^α) to φ^l, its functional logarithmic derivative shows that J takes the form $J(\varphi) = N(\varphi)\det F$, where a careful use of dimensional regularization can be applied to reduce $N(\varphi)$ to a factor depending only on the non-local functionals I^A (Liccardo 1996). Its effect is then absorbed into the measure over the field configurations, so that the $\langle \text{out} \mid \text{in} \rangle$ amplitude reads

$$\langle \text{out} \mid \text{in} \rangle = \int d\varphi \, \text{det}F \, \delta(\chi(\varphi) - \zeta) e^{iS[\varphi]}. \tag{6.B.4}$$

At this stage one performs a Gaussian average over all gauge functionals, and denoting by $\rho_{\mu\nu}$ a constant invertible matrix, and by U and V two real-valued and *independent* fermionic fields (DeWitt 1984), the path integral for the $\langle \text{out} \mid \text{in} \rangle$ amplitude is eventually re-expressed in the form

$$\langle \text{out} \mid \text{in} \rangle = \int d\varphi \, dU \, dV \, e^{i[S(\varphi) + \chi^\mu \rho_{\mu\nu} \chi^\nu + U_\alpha F^\alpha_{\ \beta} V^\beta]}. \tag{6.B.5}$$

The operator $F^\alpha_{\ \beta}$ is called the ghost operator, and the fields U and V are the corresponding ghost fields (DeWitt 1984). The physical idea is due to Feynman (1963), while the development of the corresponding formalism in quantum gravity was obtained by DeWitt (1967). However, to have a well defined path-integral representation, it is more convenient to start from a Euclidean formulation. In particular, in the one-loop quantum theory, one considers infinitesimal gauge transformations, and if the theory with action S is bosonic, one tries to choose χ^μ in such a way that both the ghost operator and the operator on perturbations of the gauge field are of Laplace type (Avramidi and Esposito 1997). Even when this is not achieved, the addition of the gauge-averaging term $\chi^\mu \rho_{\mu\nu} \chi^\nu$ plays a crucial role in ensuring that both the ghost operator and the gauge-field operator have a well defined Green's function. In the one-loop approximation, the gauge-field operator is found to have non-degenerate leading symbol (Avramidi and Esposito 1997).

The analysis of section 6.4 shows that the Eastwood–Singer gauge-averaging functional, although relying on a deep motivation at the classical level (i.e. the possibility of imposing in a conformally invariant way a supplementary condition in a theory whose field equations are, themselves, conformally invariant), spoils completely the attempt of dealing with operators of Laplace type on gauge-field perturbations. One has instead to deal with a sixth-order operator on perturbations of the potential. Once more, we would like to stress that the interpretation of this result might well be that only zeroth- or first-order derivatives of the potential can be considered in the gauge-averaging functional. On the other hand, it is quite interesting to deal with fourth-order ghost operators (see (6.4.26)), since this class of operators is already receiving careful consideration in the current literature, being part of a systematic investigation of conformally covariant operators (Branson 1996, Avramidi 1997a,c, Erdmenger 1997).

Appendix 6.C

In the analysis of heat-kernel asymptotics in section 6.6, if the three assumptions described after Eq. (6.6.19) are satisfied, one has to consider several conformal-variation formulae including the effect of the matrices Γ^i. One has then to apply the technique described in section 5.3, but it may help the general reader to see, explicitly, what one finds for some of the new invariants occurring in the coefficients a_1 and $a_{3/2}$, say. With the notation of section 5.3, the relevant formulae are as follows (cf. Avramidi and Esposito 1998a).

$$\left[\frac{d}{d\varepsilon}(K_{ij}\Gamma^i\Gamma^j)(\varepsilon)\right]_{\varepsilon=0} = -fK_{ij}\Gamma^i\Gamma^j - f_{;N}\Gamma^2, \tag{6.C.1}$$

$$\left[\frac{d}{d\varepsilon}((K_{ij}\Gamma^i\Gamma^j)^2)(\varepsilon)\right]_{\varepsilon=0} = -2f(K_{ij}\Gamma^i\Gamma^j)^2 - 2f_{;N}K_{ij}\Gamma^i\Gamma^j\Gamma^2, \tag{6.C.2}$$

$$\left[\frac{d}{d\varepsilon}(K_{ij}\Gamma^i\Gamma^j K)(\varepsilon)\right]_{\varepsilon=0} = -2fK_{ij}\Gamma^i\Gamma^j K - (m-1)f_{;N}K_{ij}\Gamma^i\Gamma^j - f_{;N}K\Gamma^2, \tag{6.C.3}$$

$$\left[\frac{d}{d\varepsilon}(K_{il}K^l{}_j\Gamma^i\Gamma^j)(\varepsilon)\right]_{\varepsilon=0} = -2fK_{il}K^l{}_j\Gamma^i\Gamma^j - 2f_{;N}K_{ij}\Gamma^i\Gamma^j, \tag{6.C.4}$$

$$\left[\frac{d}{d\varepsilon}(K_{ij}\Gamma^i\Gamma^j S)(\varepsilon)\right]_{\varepsilon=0} = -2fK_{ij}\Gamma^i\Gamma^j S + \frac{1}{2}(m-2)f_{;N}K_{ij}\Gamma^i\Gamma^j - f_{;N}S\Gamma^2, \tag{6.C.5}$$

$$\left[\frac{d}{d\varepsilon}(R_{iNjN}\Gamma^i\Gamma^j)(\varepsilon)\right]_{\varepsilon=0} = -2fR_{iNjN}\Gamma^i\Gamma^j + f_{;ij}\Gamma^i\Gamma^j$$

$$+ f_{;NN}\Gamma^2, \tag{6.C.6}$$

$$\left[\frac{d}{d\varepsilon}(R^l{}_{ilj}\Gamma^i\Gamma^j)(\varepsilon)\right]_{\varepsilon=0} = -4fR^l{}_{ilj}\Gamma^i\Gamma^j + (m-3)f_{;ij}\Gamma^i\Gamma^j$$

$$+ f_{;l}{}^{;l}\Gamma^2, \tag{6.C.7}$$

$$\left[\frac{d}{d\varepsilon}(H_{;N}K_{ij}\Gamma^i\Gamma^j)(\varepsilon)\right]_{\varepsilon=0} = -fH_{;N}K_{ij}\Gamma^i\Gamma^j - f_{;N}H_{;N}\Gamma^2. \tag{6.C.8}$$

In the integrand for the coefficient a_2, 41 new invariants occur (Avramidi and Esposito 1998a), and there is no point in listing the conformal-variation formulae for all of them. On the other hand, the interested reader might want to check some calculations. For this purpose, we present a few more equations.

$$\left[\frac{d}{d\varepsilon}(K_{ij}\Gamma^i\Gamma^jK^2)(\varepsilon)\right]_{\varepsilon=0} = -3fK_{ij}\Gamma^i\Gamma^jK^2$$

$$-2(m-1)f_{;N}K_{ij}\Gamma^i\Gamma^jK - f_{;N}K^2\Gamma^2, \tag{6.C.9}$$

$$\left[\frac{d}{d\varepsilon}(K_{ij}\Gamma^i\Gamma^jSK)(\varepsilon)\right]_{\varepsilon=0} = -3fK_{ij}\Gamma^i\Gamma^jSK$$

$$+\frac{1}{2}(m-2)f_{;N}K_{ij}\Gamma^i\Gamma^jK - (m-1)K_{ij}\Gamma^i\Gamma^jSf_{;N}$$

$$- f_{;N}SK\Gamma^2, \tag{6.C.10}$$

$$\left[\frac{d}{d\varepsilon}(R_{iNjN}\Gamma^i\Gamma^jK)(\varepsilon)\right]_{\varepsilon=0} = -3fR_{iNjN}\Gamma^i\Gamma^jK$$

$$+ f_{;ij}\Gamma^i\Gamma^jK - (m-1)f_{;N}R_{iNjN}\Gamma^i\Gamma^j$$

$$+ f_{;NN}K\Gamma^2, \tag{6.C.11}$$

$$\left[\frac{d}{d\varepsilon}(H_{;NN}K_{ij}\Gamma^i\Gamma^j)(\varepsilon)\right]_{\varepsilon=0} = -5f_{;N}H_{;N}K_{ij}\Gamma^i\Gamma^j$$

$$- fH_{;NN}K_{ij}\Gamma^i\Gamma^j - f_{;N}H_{;NN}\Gamma^2. \tag{6.C.12}$$

Appendix 6.D

The discussion in the second part of section 6.7 assumes that the reader is already familiar with the concept of moduli space, but this is not necessarily the case. Thus, some key ideas and results are presented here for completeness, relying on the monograph by Freed and Uhlenbeck (1991).

In the course of studying Yang–Mills theory, one is led to consider an orbit space, given by the set of solutions of the self-dual Yang–Mills equations, divided out by a natural equivalence. This quotient space is called the moduli space. Some basic steps are as follows. Let M be a four-dimensional Riemannian manifold, and V a vector bundle over M, with group $SU(2)$. Denoting by $\mathcal{A}$ the space of covariant derivatives

$$\mathcal{A} \equiv \{D : C^{\infty}(V) \rightarrow C^{\infty}(V \otimes T^{*}M)\}, \tag{6.D.1}$$

the Yang–Mills functional reads

$$I(D) \equiv \int_M | F_D |^2 * 1, \;\; D \in \mathcal{A}, \tag{6.D.2}$$

where F_D is the curvature of D, and $*1$ is the volume form of the metric. The Yang–Mills equations are the Euler–Lagrange equations for the action $I(D)$, and are hence obtained by setting to zero the first variation of $I(D)$:

$$\frac{d}{dt} I(D + tA) \mid_{t=0} = 2 \int_M (DA, F) * 1$$
$$= 2 \int_M (A, D^{\dagger}F) * 1, \tag{6.D.3}$$

where $D^{\dagger}$ is the formal adjoint of D, and is given by

$$D^{\dagger} = - * D*, \tag{6.D.4}$$

in terms of D and the Hodge-star operator. From what we have said, the Yang–Mills equations read

$$D^{\dagger} F_D = 0, \tag{6.D.5a}$$

or, equivalently,

$$D * F_D = 0. \tag{6.D.5b}$$

This equation is supplemented by the Bianchi identity

$$D \, F_D = 0. \tag{6.D.6}$$

The curvature F_D is then called a Yang–Mills field, and D is the corresponding Yang–Mills connection.

The curvature F can be always decomposed as the sum

$$F = F_{+} + F_{-}, \tag{6.D.7}$$

where F_{+} and F_{-} are the self-dual and anti-self-dual parts, respectively:

$$F_{+} = *F_{+}, \tag{6.D.8}$$

$$F_- = - * F_-. \tag{6.D.9}$$

If F_- vanishes, F is called self-dual; by contrast, if F_+ vanishes, F is anti-self-dual. Remarkably, if F is self-dual or anti-self-dual, the equation (6.D.5b) follows from the Bianchi identity (6.D.6), because then

$$0 = D\,F = D\,F_+ = D * F_+, \tag{6.D.10}$$

or

$$0 = D\,F = D\,F_- = -D * F_-. \tag{6.D.11}$$

Thus, self-dual or anti-self-dual gauge curvatures are, by construction, solutions of the Yang–Mills equations. The *group of gauge transformations* is denoted by $\mathcal{G}$, and is defined by

$$\mathcal{G} \equiv C^\infty(\mathrm{Aut}\,V). \tag{6.D.12}$$

The *moduli space* is then defined as the quotient space

$$\mathcal{M} \equiv \{D \in \mathcal{A} : F_D = *F_D\}/\mathcal{G}. \tag{6.D.13}$$

The topology of the moduli space is a fascinating mathematical subject, but before we can state the main theorem about this issue, we have to define another useful mathematical structure. We are here interested in the *intersection form*, which is the basic invariant of compact four-manifolds. It can be defined in one of the following three ways.

(i) Let M be a compact, oriented, simply connected and topological four-manifold. A symmetric bilinear form on $H_2(M;Z)$ can be obtained by considering two elements of $H_2(M;Z)$, say a and b, represented by oriented surfaces A, B in M which intersect transversely. One then defines

$$\omega(a,b) \equiv A \cdot B, \tag{6.D.14}$$

where $A \cdot B$ is the intersection number of the cycles A and B. By construction, ω is symmetric, since both A and B are two-dimensional. Moreover, ω is unimodular, i.e. any matrix chosen to represent ω has determinant ± 1.

(ii) By virtue of Poincaré duality, one can identify $H_2(M;Z)$ with $H^2(M;Z)$, and hence ω can be defined as a pairing

$$\omega : H^2(M;Z) \otimes H^2(M;Z) \rightarrow Z.$$

If α, β are both elements of $H^2(M; Z)$, and $[M]$ of $H_4(M; Z)$ is the orientation, one defines

$$\omega(\alpha, \beta) \equiv \Bigl(\alpha \cup \beta\Bigr)[M]. \tag{6.D.15}$$

With our notation, the map

$$\cup : H^2(M; Z) \otimes H^2(M; Z) \rightarrow H^4(M; Z)$$

is the cup-product (see section 2.3).

(iii) If the manifold M is smooth, one may define ω with the help of the de Rham cohomology $H^*_{DR}(M)$. For this purpose, if α, β are elements of $H^2_{DR}(M)$, represented by two-forms denoted again (for simplicity) by α and β, the definition of ω requires that

$$\omega(\alpha, \beta) \equiv \int_M \alpha \wedge \beta. \tag{6.D.16}$$

Now we can, at last, state a key theorem in the theory of the topology of the moduli space (Freed and Uhlenbeck 1991). The mathematical framework is given by an $SU(2)$ vector bundle over a compact, simply connected, oriented smooth four-manifold M, whose intersection form is positive-definite. First, one studies the equation

$$\omega(\alpha, \alpha) = 1. \tag{6.D.17}$$

If n is half the number of solutions of Eq. (6.D.17), then for almost all metrics on M, there exist 'points' $p_1, ..., p_n$ of the moduli space $\mathcal{M}$ such that $\mathcal{M} - \{p_1, ..., p_n\}$ is a smooth five-manifold. Such points are in one-to-one correspondence with topological splittings of the vector bundle V. Second, one finds that neighbourhoods, say $\mathcal{O}_{p_i}$, of the singular points p_i exist such that

$$\mathcal{O}_{p_i} \cong \text{cone on } CP^2. \tag{6.D.18}$$

Third, orientability of the moduli space is proved. Fourth, the set $\mathcal{M} - \{p_1, ..., p_n\}$ is always non-empty. Fifth, the set $\overline{\mathcal{M}}$ is a compact smooth manifold with boundary.

The proof of these properties is the result of the dedicated work by some of the leading mathematicians of the end of the twentieth century. Their statement may be taken as the starting point, if the reader wants to understand them thoroughly, and possibly apply similar techniques to the moduli space of the Seiberg–Witten equations (Donaldson 1996), that we have mentioned in section 6.7.

References

Alexandrov, S. & Vassilevich, D. V. (1996). Heat kernel for non-minimal operators on a Kähler manifold. *J. Math. Phys.*, **37**, 5715–5718.

Alvarez-Gaumé, L. (1983a). Supersymmetry and the Atiyah–Singer index theorem. *Commun. Math. Phys.*, **90**, 161–173.

Alvarez-Gaumé, L. (1983b). A Note on the Atiyah–Singer index theorem. *J. Phys.*, **A 16**, 4177–4182.

Amsterdamski, P., Berkin, A. & O'Connor, D. (1989). b_8 Hamidew coefficient for a scalar field. *Class. Quantum Grav.*, **6**, 1981–1991.

Ashtekar, A. (1988). *New Perspectives in Canonical Gravity.* Naples: Bibliopolis.

Ashtekar, A. (1991). *Lectures on Non-Perturbative Canonical Gravity.* Singapore: World Scientific.

Atiyah, M. F. & Singer, I. M. (1963). The index of elliptic operators on compact manifolds. *Bull. Amer. Math. Soc.*, **69**, 422–433.

Atiyah, M. F. & Bott, R. (1964). On the periodicity theorem for complex vector bundles. *Acta Math.*, **112**, 229–247.

Atiyah, M. F. & Bott, R. H. (1965). The index problem for manifolds with boundary. In *Differential Analysis*, papers presented at the Bombay Colloquium 1964, pp. 175–186. Oxford: Oxford University Press.

Atiyah, M. F. (1966). Global aspects of the theory of elliptic differential operators. In *Proc. Int. Congress of Mathematicians*, Moscow, pp. 57–64.

Atiyah, M. F. (1967). *K-Theory.* New York: Benjamin.

Atiyah, M. F. & Singer, I. M. (1968a). The index of elliptic operators: I. *Ann. of Math.*, **87**, 484–530.

Atiyah, M. F. & Segal, G. B. (1968). The index of elliptic operators: II. *Ann. of Math.* **87**, 531–545.

Atiyah, M. F. & Singer, I. M. (1968b). The index of elliptic operators: III. *Ann. of Math.*, **87**, 546–604.

Atiyah, M. F. & Singer, I. M. (1971a). The index of elliptic operators: IV. *Ann. of Math.*, **93**, 119–138.
Atiyah, M. F. & Singer, I. M. (1971b). The index of elliptic operators: V. *Ann. of Math.*, **93**, 139–149.
Atiyah, M. F. & Patodi, V. K. (1973). On the heat equation and the index theorem. *Inventiones Math.*, **19**, 279–330 [Erratum in *Inventiones Math.*, **28**, 277–280].
Atiyah, M. F. (1975a). Classical groups and classical differential operators on manifolds. In *Differential Operators on Manifolds*, CIME, Varenna 1975, pp. 5–48. Roma: Edizione Cremonese.
Atiyah, M. F. (1975b). Eigenvalues and Riemannian geometry. In *Proc. Int. Conf. on Manifolds and Related Topics in Topology*, pp. 5–9. Tokyo: University of Tokyo Press.
Atiyah, M. F., Patodi, V. K. & Singer, I. M. (1975). Spectral asymmetry and Riemannian geometry I. *Math. Proc. Camb. Phil. Soc.*, **77**, 43–69.
Atiyah, M. F., Patodi, V. K. & Singer, I. M. (1976). Spectral asymmetry and Riemannian geometry III. *Math. Proc. Camb. Phil. Soc.*, **79**, 71–99.
Atiyah, M. F. (1984a). Topological aspects of anomalies. In *Symposium on Anomalies, Geometry, Topology*, pp. 22–32. Singapore: World Scientific.
Atiyah, M. F. (1984b). Anomalies and index theory. In *Lecture Notes in Physics*, Vol. **208**, pp. 313–322. Berlin: Springer-Verlag.
Atiyah, M. F. & Singer, I. M. (1984). Dirac operators coupled to vector potentials. *Proc. Nat. Acad. of Sci. USA*, **81**, 2597–2600.
Atiyah, M. F. (1985). Eigenvalues of the Dirac operator. In *Proc. of 25th Mathematics Arbeitstagung, Bonn 1984, Lecture Notes in Mathematics*, pp. 251–260. Berlin: Springer-Verlag.
Avis, S. J. & Isham, C. J. (1980). Generalized spin-structures on four-dimensional space-times. *Commun. Math. Phys.*, **72**, 103–118.
Avramidi, I. G. (1989). Background field calculations in quantum field theory (vacuum polarization). *Theor. Math. Phys.*, **79**, 494–502.
Avramidi, I. G. (1990a). The non-local structure of one-loop effective action via partial summation of asymptotic expansion. *Phys. Lett.*, **B 236**, 443–449.
Avramidi, I. G. (1990b). The covariant technique for calculation of the heat-kernel asymptotic expansion. *Phys. Lett.*, **B 238**, 92–97.
Avramidi, I. G. (1991). A covariant technique for the calculation of the one-loop effective action. *Nucl. Phys.*, **B 355**, 712–754.
Avramidi, I. G. & Schimming, R. (1996). Algorithms for the calculation of the heat-kernel coefficients. In *Quantum Field Theory*

Under the Influence of External Conditions, ed. M. Bordag, pp. 150–163. Leipzig: Teubner.

Avramidi, I. G., Esposito, G. & Kamenshchik, A. Yu. (1996). Boundary operators in Euclidean quantum gravity. *Class. Quantum Grav.*, **13**, 2361–2373.

Avramidi, I. G. (1997a). Singularities of Green functions of the products of the Laplace type operators. *Phys. Lett.*, **B 403**, 280–284.

Avramidi, I. G. (1997b). Covariant techniques for computation of the heat kernel (HEP-TH 9704166, extended version of a lecture given at the University of Iowa).

Avramidi, I. G. (1997c). Green functions of higher order differential operators (HEP-TH 9707040).

Avramidi, I. G. & Esposito, G. (1997). Gauge theories on manifolds with boundary (HEP-TH 9710048).

Avramidi, I. G. & Esposito, G. (1998a). New invariants in the one-loop divergences on manifolds with boundary. *Class. Quantum Grav.*, **15**, 281–297.

Avramidi, I. G. & Esposito, G. (1998b). Lack of strong ellipticity in Euclidean quantum gravity. *Class. Quantum Grav.*, **15**, 1141–1152.

Bär, C. (1992a). Upper eigenvalue estimates for Dirac operators. *Ann. Glob. Anal. Geom.*, **10**, 171–177.

Bär, C. & Schmutz, P. (1992). Harmonic spinors on Riemann surfaces. *Ann. Glob. Anal. Geom.*, **10**, 263–273.

Bär, C. (1992b). The Dirac fundamental tone of the hyperbolic space. *Geometriae Dedicata*, **41**, 103–107.

Bär, C. (1992c). The Dirac operator on homogeneous spaces and its spectrum on three-dimensional lens spaces. *Arch. Math.*, **59**, 65–79.

Bär, C. (1992d). Lower eigenvalue estimates for Dirac operators. *Math. Ann.*, **293**, 39–46.

Bär, C. (1995). Harmonic spinors for twisted Dirac operators (Sfb 288 Preprint No. 180).

Bär, C. (1996a). Metrics with harmonic spinors. *Geom. Func. Anal.*, **6**, 899–942.

Bär, C. (1996b). The Dirac operator on space forms of positive curvature. *J. Math. Soc. Japan*, **48**, 69–83.

Bär, C. (1996c). On nodal sets for Dirac and Laplace operators (Freiburg Preprint 29/1996).

Bär, C. (1997). Harmonic spinors and topology (Freiburg Preprint 6/1997).

Barvinsky, A. O. (1987). The wave function and the effective action in quantum cosmology: covariant loop expansion. *Phys. Lett.*, **B 195**, 344–348.

Barvinsky, A. O., Kamenshchik, A. Yu. & Karmazin, I. P. (1992). One-loop quantum cosmology: ζ-function technique for the Hartle-Hawking wave function of the universe. *Ann. Phys.*, **219**, 201–242.

Barvinsky, A. O., Gusev, Yu. V., Vilkovisky, G. A. & Zhytnikov, V. V. (1994). Asymptotic behaviours of the heat kernel in covariant perturbation theory. *J. Math. Phys.*, **35**, 3543–3559.

Baston, R. J. & Eastwood, M. (1990). Invariant operators. In *Twistors in Mathematics and Physics*, L.M.S. Lecture Notes **156**, eds. T. N. Bailey and R. J. Baston, pp. 129–163. Cambridge: Cambridge University Press.

Bellisai, D. (1996). Fermionic zero-modes around string solitons. *Nucl. Phys.*, **B 467**, 127–145.

Bérard, P. H. (1986). *Spectral Geometry: Direct and Inverse Problems.* Lecture Notes in Mathematics, Vol. 1207. Berlin: Springer-Verlag.

Berline, N., Getzler, E. & Vergne, M. (1992). *Heat Kernels and Dirac Operators.* Berlin: Springer-Verlag.

Berry, M. V. & Mondragon, R. J. (1987). Neutrino billiards: time-reversal symmetry-breaking without magnetic fields. *Proc. R. Soc. Lond.*, **A 412**, 53–74.

Bismut, J. M. (1986a). Localization formulas, superconnections, and the index theorem for families. *Commun. Math. Phys.*, **103**, 127–166.

Bismut, J. M. (1986b). The Atiyah–Singer index theorem for families of Dirac operators: two heat-equation proofs. *Invent. Math.*, **83**, 91–151.

Booss, B. & Bleecker, D. D. (1985). *Topology and Analysis: The Atiyah–Singer Index Formula and Gauge-Theoretic Physics.* Berlin: Springer-Verlag.

Booss-Bavnbek, B. & Wojciechowski, K. P. (1993). *Elliptic Boundary Problems for Dirac Operators.* Boston: Birkhäuser.

Booss-Bavnbek, B., Morchio, G., Strocchi, F. & Wojciechowski, K. P. (1997). Grassmannian and Chiral Anomaly. *J. Geom. Phys.*, **22**, 219–244.

Bordag, M., Elizalde, E. & Kirsten, K. (1996a). Heat-kernel coefficients of the Laplace operator on the d-dimensional ball. *J. Math. Phys.*, **37**, 895–916.

Bordag, M., Geyer, B., Kirsten, K. & Elizalde, E. (1996b). Zeta-function determinant of the Laplace operator on the d-dimensional ball. *Commun. Math. Phys.*, **179**, 215–234.

Bordag, M., Kirsten, K. & Dowker, J. S. (1996c). Heat kernels and functional determinants on the generalized cone. *Commun. Math.*

Phys., **182**, 371–394.

Bott, R. (1959). The stable homotopy of the classical groups. *Ann. of Math.*, **70**, 313–337.

Boyer, T. H. (1968). Quantum electromagnetic zero-point energy of a conducting spherical shell and the Casimir model for a charged particle. *Phys. Rev.*, **174**, 1764–1776.

Branson, T. P., Gilkey, P. B. & Orsted, B. (1990). Leading terms in the heat invariants. *Proc. Amer. Math. Soc.*, **109**, 437–450.

Branson, T. P. & Gilkey, P. B. (1990). The asymptotics of the Laplacian on a manifold with boundary. *Commun. in Partial Differential Equations*, **15**, 245–272.

Branson, T. P. & Gilkey, P. B. (1992a). Residues of the η-function for an operator of Dirac type. *J. Funct. Anal.*, **108**, 47–87.

Branson, T. P. & Gilkey, P. B. (1992b). Residues of the η-function for an operator of Dirac type with local boundary conditions. *Diff. Geom. Appl.*, **2**, 249–267.

Branson, T. P. & Gilkey, P. B. (1994). Functional determinant of a four-dimensional boundary-value problem. *Trans. Amer. Math. Soc.*, **344**, 479–531.

Branson, T. P. (1995). Sharp inequalities, the functional determinant, and the complementary series. *Trans. A. M. S.*, **347**, 3671–3742.

Branson, T. P. (1996). An anomaly associated with four-dimensional quantum gravity. *Commun. Math. Phys.*, **178**, 301–309.

Branson, T. P., Gilkey, P. B. & Vassilevich, D. V. (1997). Vacuum expectation value asymptotics for second-order differential operators on manifolds with boundary (HEP-TH 9702178).

Breitenlohner, P. & Freedman, D. Z. (1982). Stability in gauged extended supergravity. *Ann. Phys.*, **144**, 249–281.

Camporesi, R. & Higuchi, A. (1996). On the eigenfunctions of the Dirac operator on spheres and real hyperbolic spaces. *J. Geom. Phys.*, **20**, 1–18.

Carow-Watamura, U. & Watamura, S. (1997). Chirality and Dirac operator on non-commutative sphere. *Commun. Math. Phys.*, **183**, 365–382.

Carslaw, H. S. & Jaeger, J. C. (1959). *Conduction of Heat in Solids.* Oxford: Clarendon Press.

Casimir, H. B. G. (1948). On the attraction between two perfectly conducting plates. *Proc. Kon. Ned. Akad. Wetenschap.*, ser. B, **51**, 793–795.

Chavel, I. (1984). *Eigenvalues in Riemannian Geometry.* New York: Academic Press.

Chern, S. S. (1944). A simple intrinsic proof of the Gauss–Bonnet formula for closed Riemannian manifolds. *Ann. of Math.*, **45**, 741–752.

Chern, S. S. (1945). On the curvatura integra in a Riemannian manifold. *Ann. of Math.*, **46**, 674–684.

Chern, S. S. (1979). *Complex Manifolds Without Potential Theory.* Berlin: Springer-Verlag.

Chernoff, P. R. (1977). Schrödinger and Dirac operators with singular potentials and hyperbolic equations. *Pacific J. Math.*, **72**, 361–382.

Chodos, A., Jaffe, R. L., Johnson, K., Thorn, C. B. & Weisskopf, V. F. (1974). New extended model of hadrons. *Phys. Rev.*, **D 9**, 3471–3495.

Cognola, G. and Zerbini, S. (1988). Seeley–DeWitt coefficients in a Riemann–Cartan manifold. *Phys. Lett.*, **B 214**, 70–74.

Connes, A. (1995). Non-commutative geometry and physics. In *Gravitation and Quantizations, Les Houches Session LVII*, eds. B. Julia and J. Zinn-Justin, pp. 805–950. Amsterdam: Elsevier.

Dabrowski, L. & Trautman, A. (1986). Spinor structures on spheres and projective spaces. *J. Math. Phys.*, **27**, 2022–2028.

Davies, E. B. (1989). *Heat Kernels and Spectral Theory.* Cambridge: Cambridge University Press.

Davies, E. B. (1997). L^p spectral theory of higher-order elliptic differential operators. *Bull. London Math. Soc.*, **29**, 513–546.

D'Eath, P. D. (1984). Canonical quantization of supergravity. *Phys. Rev.*, **D 29**, 2199–2219.

D'Eath, P. D. & Halliwell, J. J. (1987). Fermions in quantum cosmology. *Phys. Rev.*, **D 35**, 1100–1123.

D'Eath, P. D. & Esposito, G. (1991a). Local boundary conditions for the Dirac operator and one-loop quantum cosmology. *Phys. Rev.*, **D 43**, 3234–3248.

D'Eath, P. D. & Esposito, G. (1991b). Spectral boundary conditions in one-loop quantum cosmology. *Phys. Rev.*, **D 44**, 1713–1721.

D'Eath, P. D. (1996). *Supersymmetric Quantum Cosmology.* Cambridge: Cambridge University Press.

De Nardo, L., Fursaev, D. V. & Miele, G. (1997). Heat-kernel coefficients and spectra of the vector Laplacians on spherical domains with conical singularities. *Class. Quantum Grav.*, **14**, 1059–1078.

DeWitt, B. S. (1965). *Dynamical Theory of Groups and Fields.* New York: Gordon and Breach.

DeWitt, B. S. (1967). Quantum theory of gravity. II. The manifestly covariant theory. *Phys. Rev.*, **162**, 1195–1239.

DeWitt, B. S. (1981). A gauge-invariant effective action. In *Quantum*

Gravity 2, A Second Oxford Symposium, eds. C. J. Isham, R. Penrose and D. W. Sciama, pp. 449–487. Oxford: Clarendon Press.

DeWitt, B. S. (1984). The space-time approach to quantum field theory. In *Relativity, Groups and Topology II*, eds. B. S. DeWitt and R. Stora, pp. 381–738. Amsterdam: North Holland.

Dirac, P. A. M. (1928). The quantum theory of the electron. *Proc. R. Soc. Lond.*, **A 117**, 610–624.

Dirac, P. A. M. (1958). *The Principles of Quantum Mechanics.* Oxford: Clarendon Press.

Donaldson, S. K. & Kronheimer, P. B. (1990). *The Geometry of Four-Manifolds.* Oxford: Clarendon Press.

Donaldson, S. (1996). The Seiberg–Witten equations and 4-manifold topology. *Bull. Am. Math. Soc.*, **33**, 45–70.

Dowker, J. S. & Schofield, J. P. (1990). Conformal transformations and the effective action in the presence of boundaries. *J. Math. Phys.*, **31**, 808–818.

Dowker, J. S. & Apps, J. S. (1995). Further functional determinants. *Class. Quantum Grav.*, **12**, 1363–1383.

Dowker, J. S. (1996a). Spin on the four-ball. *Phys. Lett.*, **B 366**, 89–94.

Dowker, J. S. (1996b). Robin conditions on the Euclidean ball. *Class. Quantum Grav.*, **13**, 585–610.

Dowker, J. S., Apps, J. S., Kirsten, K. & Bordag, M. (1996). Spectral invariants for the Dirac equation on the d-ball with various boundary conditions. *Class. Quantum Grav.*, **13**, 2911–2920.

Dowker, J. S. & Kirsten, K. (1996). Spinors and forms on generalized cones (HEP-TH 9608189).

Dowker, J. S. & Kirsten, K. (1997). Heat-kernel coefficients for oblique boundary conditions. *Class. Quantum Grav.*, **14**, L169–L175.

Dürr, S. & Wipf, A. (1997). Finite temperature Schwinger model with chirality breaking boundary conditions. *Ann. Phys.*, **255**, 333–361.

Eastwood, M. & Singer, I. M. (1985). A conformally invariant Maxwell gauge. *Phys. Lett.*, **A 107**, 73–74.

Eastwood, M. & Rice, J. (1987). Conformally invariant differential operators on Minkowski space and their curved analogues. *Commun. Math. Phys.*, **109**, 207–228 [Erratum in *Commun. Math. Phys.*, **144**, 213].

Eisenhart, L. P. (1926). *Riemannian Geometry.* Princeton: Princeton University Press.

Elizalde, E. (1995). *Ten Physical Applications of Spectral Zeta-Functions.* Lecture Notes in Physics, New Series m: Monographs, Vol. m35. Berlin: Springer-Verlag.

Endo, R. (1995). Heat kernel for spin-$\frac{3}{2}$ Rarita–Schwinger field in general covariant gauge. *Class. Quantum Grav.*, **12**, 1157–1164.

Erdmenger, J. (1997). Conformally covariant operators: theory and applications. *Class. Quantum Grav.*, **14**, 2061–2084.

Esposito, G. (1994a). *Quantum Gravity, Quantum Cosmology and Lorentzian Geometries.* Lecture Notes in Physics, New Series m: Monographs, Vol. m12, second corrected and enlarged edition. Berlin: Springer-Verlag.

Esposito, G. (1994b). Gauge-averaging functionals for Euclidean Maxwell theory in the presence of boundaries. *Class. Quantum Grav.*, **11**, 905–926.

Esposito, G. & Kamenshchik, A. Yu. (1994). Coulomb gauge in one-loop quantum cosmology. *Phys. Lett.*, **B 336**, 324–329.

Esposito, G., Kamenshchik, A. Yu., Mishakov, I. V. & Pollifrone, G. (1994a). Euclidean Maxwell theory in the presence of noundaries. II. *Class. Quantum Grav.*, **11**, 2939–2950.

Esposito, G., Kamenshchik, A. Yu., Mishakov, I. V. & Pollifrone, G. (1994b). Gravitons in one-loop quantum cosmology. Correspondence between covariant and non-covariant formalisms. *Phys. Rev.*, **D 50**, 6329–6337.

Esposito, G. (1995). *Complex General Relativity.* Fundamental Theories of Physics, Vol. 69. Dordrecht: Kluwer.

Esposito, G., Kamenshchik, A. Yu., Mishakov, I. V. & Pollifrone, G. (1995a). Relativistic gauge conditions in quantum cosmology. *Phys. Rev.*, **D 52**, 2183–2191.

Esposito, G., Kamenshchik, A. Yu., Mishakov, I. V. & Pollifrone, G. (1995b). One-loop amplitudes in Euclidean quantum gravity. *Phys. Rev.*, **D 52**, 3457–3465.

Esposito, G. & Kamenshchik, A. Yu. (1995). Mixed boundary conditions in Euclidean quantum gravity. *Class. Quantum Grav.*, **12**, 2715–2722.

Esposito, G. & Kamenshchik, A. Yu. (1996). One-loop divergences in simple supergravity: boundary effects. *Phys. Rev.*, **D 54**, 3869–3881.

Esposito, G., Morales-Técotl, H. A. & Pimentel, L. O. (1996). Essential self-adjointness in one-loop quantum cosmology. *Class. Quantum Grav.*, **13**, 957–963.

Esposito, G. (1996). Local boundary conditions in quantum supergravity. *Phys. Lett.*, **B 389**, 510–514.

Esposito, G. (1997a). Non-local boundary conditions for massless spin-$\frac{1}{2}$ fields. *Phys. Rev.*, **D 55**, 3886–3888.

Esposito, G. (1997b). Quantized Maxwell theory in a conformally invariant gauge. *Phys. Rev.*, **D 56**, 2442–2444.

Esposito, G., Kamenshchik, A. Yu. & Pollifrone, G. (1997). *Euclidean Quantum Gravity on Manifolds with Boundary.* Fundamental Theories of Physics, Vol. 85. Dordrecht: Kluwer.
Estrada, R. & Kanwal, R. P. (1990). A distributional theory for asymptotic expansions. *Proc. Roy. Soc. London*, **A 428**, 399–430.
Estrada, R. & Kanwal, R. P. (1994). *Asymptotic Analysis: A Distributional Approach.* Boston: Birkhäuser.
Estrada, R. & Fulling, S. A. (1997). Distributional asymptotic expansions of spectral functions and of the associated Green kernels (FUNCT-AN 9710003).
Faddeev, L. D. & Popov, V. (1967). Feynman diagrams for the Yang–Mills field. *Phys. Lett.*, **B 25**, 29–30.
Falomir, H., Gamboa Saraví R. E., Muschietti, M. A., Santangelo, E. M. & Solomin, J. E. (1996a). Determinants of Dirac operators with local boundary conditions. *J. Math. Phys.*, **37**, 5805–5819.
Falomir, H., Gamboa Saraví R. E. & Santangelo, E. M. (1996b). Dirac operator on a disk with global boundary conditions (HEP-TH 9609194).
Falomir, H. (1996). Condiciones de contorno globales para el operator de Dirac (HEP-TH 9612155).
Falomir, H. (1997). Global boundary conditions for the Dirac operator (PHYSICS 9705013, talk given at the *Trends in Theoretical Physics, CERN-Santiago de Compostela-La Plata Meeting*).
Feynman, R. P. (1963). Quantum theory of gravitation. *Acta Physica Polonica*, **24**, 697–722.
Fleckinger, J., Levitin, M. & Vassiliev, D. (1995). Heat equation on the triadic von Koch snowflake: asymptotic and numerical analysis. *Proc. London Math. Soc.*, **71**, 372–396.
Fock, V. A. (1937). The proper time in classical and quantum mechanics. *Izvestiya of USSR Academy of Sciences (Phys.)*, **4–5**, 551–568.
Freed, D. S. & Uhlenbeck, K. K. (1991). *Instantons and Four-Manifolds, Second Edition.* Mathematical Sciences Research Institute Publications, Vol. 1. New York: Springer-Verlag.
Freed, D. S. & Uhlenbeck, K. K. (1995). *Geometry and Quantum Field Theory.* American Mathematical Society.
Fulling, S. A. (1989). *Aspects of Quantum Field Theory in Curved Space-Time.* Cambridge: Cambridge University Press.
Fulling, S. A. (1995). *Heat Kernel Techniques and Quantum Gravity (Discourses in Mathematics and Its Applications, No. 4).* College Station: Texas A&M University.
Fulling, S. A. (1996). Pseudodifferential operators, covariant

quantization, the inescapable VanVleck–Morette determinant, and the $\frac{R}{6}$ controversy. *Int. J. Mod. Phys.*, **D 5**, 597–608.
Fursaev, D. V. (1994). Spectral geometry and one-loop divergences on manifolds with conical singularities. *Phys. Lett.*, **B 334**, 53–60.
Fursaev, D. V. & Miele, G. (1997). Cones, spins and heat kernels. *Nucl. Phys.*, **B 484**, 697–723.
Gibbons, G. W. & Hawking, S. W. (1977). Action integrals and partition functions in quantum gravity. *Phys. Rev.*, **D 15**, 2752–2756.
Gibbons, G. W. & Hawking, S. W. (1993). *Euclidean Quantum Gravity.* Singapore: World Scientific.
Gilkey, P. B. (1975). The spectral geometry of a Riemannian manifold. *J. Diff. Geom.*, **10**, 601–618.
Gilkey, P. B. (1995). *Invariance Theory, the Heat Equation, and the Atiyah–Singer Index Theorem.* Boca-Raton: Chemical Rubber Company.
Goldthorpe, W. H. (1980). Spectral geometry and SO(4) gravity in a Riemann Cartan space-time. *Nucl. Phys.*, **B 170**, 307–328.
Graham, C. R., Jenne, R., Mason, L. J. & Sparling, G. A. J. (1992). Conformally invariant powers of the Laplacian. I: existence. *J. Lond. Math. Soc.*, **46**, 557–565.
Greiner, P. (1971). An asymptotic expansion for the heat equation. *Archs. Ration. Mech. Analysis*, **41**, 163–218.
Greub, W., Halperin, S. & Vanstone, R. (1973). *Connections, Curvature and Cohomology*, Vol. II. New York: Academic.
Grib, A. A., Mamaev, S. G. & Mostepanenko, V. M. (1994). *Vacuum Quantum Effects in Strong Fields.* St. Petersburg: Friedmann Laboratory Publishing.
Grubb, G. & Seeley, R. T. (1995). Weakly parametric pseudodifferential operators and Atiyah–Patodi–Singer boundary problems. *Inv. Math.*, **121**, 481–529.
Grubb, G. (1996). *Functional Calculus of Pseudodifferential Boundary Problems.* Progress in Mathematics, Vol. 65. Boston: Birkhäuser.
Günther, P. & Schimming, R. (1977). Curvature and spectrum of compact Riemannian manifolds. *J. Diff. Geom.*, **12**, 599–618.
Gusynin, V. P., Gorbar, E. V. & Romankov, V. V. (1991). Heat-kernel expansions for non-minimal differential operators and manifolds with torsion. *Nucl. Phys.*, **B 362**, 449–471.
Hadamard, J. (1952). *Lectures on Cauchy's Problem in Linear Differential Equations.* New York: Dover.
Hartle, J. B. & Hawking, S. W. (1983). Wave function of the universe. *Phys. Rev.*, **D 28**, 2960–2975.

Hawking, S. W. & Ellis, G. F. R. (1973). *The Large-Scale Structure of Space-Time.* Cambridge: Cambridge University Press.

Hawking, S. W. (1977). Zeta-function regularization of path integrals in curved space-time. *Commun. Math. Phys.*, **55**, 133–148.

Hawking, S. W. (1979). The path-integral approach to quantum gravity. In *General Relativity, an Einstein Centenary Survey*, eds. S. W. Hawking and W. Israel, pp. 746–789. Cambridge: Cambridge University Press.

Hawking, S. W. (1982). The boundary conditions of the universe. In *Pontificiae Academiae Scientiarum Scripta Varia*, **48**, 563–574.

Hawking, S. W. (1983). The boundary conditions for gauged supergravity. *Phys. Lett.*, **B 126**, 175–177.

Hawking, S. W. (1984). The quantum state of the universe. *Nucl. Phys.*, **B 239**, 257–276.

Hawking, S. W. (1996). Virtual black holes. *Phys. Rev.*, **D 53**, 3099–3107.

Hitchin, N. (1974). Harmonic spinors. *Adv. in Math.*, **14**, 1–55.

Hörmander, L. (1963). *Linear Partial Differential Operators.* Berlin: Springer-Verlag.

Hortacsu, M., Rotke, K. D. & Schroer, B. (1980). Zero-energy eigenstates for the Dirac boundary problem. *Nucl. Phys.*, **B 171**, 530–542.

Immirzi, G. (1997). Real and complex connections for canonical gravity. *Class. Quantum Grav.*, **14**, L177–L181.

Isham, C. J. (1978). Spinor fields in four-dimensional space-time. *Proc. R. Soc. Lond.*, **A 364**, 591–599.

Isham, C. J. (1989). *Modern Differential Geometry for Physicists.* Singapore: World Scientific.

Itzykson, C. & Zuber J. B. (1985). *Quantum Field Theory.* Singapore: McGraw-Hill.

Johnson, K. (1975). The M. I. T. bag model. *Acta Phys. Polonica*, **B 6**, 865–892.

Jona-Lasinio, G. (1964). Relativistic field theories with symmetry-breaking solutions. *Nuovo Cimento*, **34**, 1790–1795.

Kac, M. (1966). Can one hear the shape of a drum? *Amer. Math. Monthly*, **73**, 1–23.

Kamenshchik, A. Yu. & Mishakov, I. V. (1992). ζ-function technique for quantum cosmology: the contributions of matter fields to the Hartle–Hawking wave function of the universe. *Int. J. Mod. Phys.*, **A 7**, 3713–3746.

Kamenshchik, A. Yu. & Mishakov, I. V. (1993). Fermions in one-loop quantum cosmology. *Phys. Rev.*, **D 47**, 1380–1390.

Kamenshchik, A. Yu. & Mishakov, I. V. (1994). Fermions in one-loop quantum cosmology II. The problem of correspondence between covariant and non-covariant formalisms. *Phys. Rev.*, **D 49**, 816–824.

Kellogg, O. D. (1954). *Foundations of Potential Theory.* New York: Dover.

Kennedy, G. (1978). Boundary terms in the Schwinger–DeWitt expansion: flat-space results. *J. Phys.*, **A 11**, L173–L178.

Kirsten, K. & Cognola, G. (1996). Heat-kernel coefficients and functional determinants for higher-spin fields on the ball. *Class. Quantum Grav.*, **13**, 633–644.

Kirsten, K. (1998). The a_5 coefficient on a manifold with boundary. *Class. Quantum Grav.*, **15**, L5–L12.

Kori, T. (1996). Index of the Dirac operator on S^4 and the infinite dimensional Grassmannian on S^3. *Japan J. Math.*, **22**, 1–36.

Lamoreaux, S. K. (1997). Demonstration of the Casimir force in the 0.6 to 6 μm range. *Phys. Rev. Lett.*, **78**, 5–8.

Landi, G. & Rovelli, C. (1997). General relativity in terms of Dirac eigenvalues. *Phys. Rev. Lett.*, **78**, 3051–3054.

LeBrun, C. (1995). Einstein metrics and Mostov rigidity. *Math. Res. Letters*, **2**, 1–8.

Levitin, M. (1998). Dirichlet and Neumann heat invariants for Euclidean Balls. *Diff. Geom. and Its Appl.*, **8**, 35–46.

Liccardo, A. (1996). Azione effettiva alla DeWitt e tecniche di calcolo in teorie con bordo. Graduation thesis, University of Naples.

Lichnerowicz, A. (1963). Spineurs harmoniques. *C. R. Acad. Sci. Paris Ser. A-B*, **257**, 7–9.

Luckock, H. C. & Moss, I. G. (1989). The quantum geometry of random surfaces and spinning membranes. *Class. Quantum Grav.*, **6**, 1993–2027.

Luckock, H. C. (1991). Mixed boundary conditions in quantum field theory. *J. Math. Phys.*, **32**, 1755–1766.

Magnus, W., Oberhettinger, F. & Soni, R. P. (1966). *Formulas and Theorems for the Special Functions of Mathematical Physics.* Berlin: Springer-Verlag.

Marachevsky, V. N. & Vassilevich, D. V. (1996). A diffeomorphism-invariant eigenvalue problem for metric perturbations in a bounded region. *Class. Quantum Grav.*, **13**, 645–652.

Martellini, M. & Reina, C. (1985) Some remarks on the index theorem approach to anomalies. *Ann. Inst. Henri Poincaré*, **43**, 443–458.

McAvity, D. M. & Osborn, H. (1991a). A DeWitt expansion of the heat kernel for manifolds with a boundary. *Class. Quantum Grav.*,

8, 603–638.
McAvity, D. M. & Osborn, H. (1991b). Asymptotic expansion of the heat kernel for generalized boundary conditions. *Class. Quantum Grav.*, **8**, 1445–1454.
McAvity, D. M. (1992). Heat-kernel asymptotics for mixed boundary conditions. *Class. Quantum Grav.*, **9**, 1983–1998.
McKean, H. P. & Singer, I. M. (1967). Curvature and eigenvalues of the Laplacian. *J. Diff. Geom.*, **1**, 43–69.
Milnor, J. W. (1963a). *Morse Theory.* Princeton: Princeton University Press.
Milnor, J. W. (1963b). Spin-structures on manifolds. *Enseign. Math.*, **9**, 198–203.
Milnor, J. W. & Stasheff, J. D. (1974). *Characteristic Classes.* Princeton: Princeton University Press.
Minakshisundaram, S. & Pleijel, A. (1949). Some properties of the eigenfunctions of the Laplace operator on Riemannian manifolds. *Can. J. Math.*, **1**, 242–256.
Minakshisundaram, S. (1953). Eigenfunctions on Riemannian manifolds. *J. Indian Math. Soc.*, **17**, 158–165.
Mishchenko, A. V. & Sitenko, A. Yu. (1992). Spectral boundary conditions and index theorem for two-dimensional compact manifold with boundary. *Ann. Phys.*, **218**, 199–232.
Mishchenko, A. V. & Sitenko, A. Yu. (1993). Zero modes of the two-dimensional Dirac operator on a noncompact Riemann surface in an external magnetic field. *Phys. At. Nucl.*, **56**, 131–137.
Moniz, P. (1996). Supersymmetric quantum cosmology: shaken not stirred. *Int. J. Mod. Phys.*, **A 11**, 4321–4382.
Morgan, J. W. (1996). *The Seiberg–Witten Equations and Applications to the Topology of Smooth Four-Manifolds.* Mathematical Notes **44**. Princeton: Princeton University Press.
Moss, I. G. (1989). Boundary terms in the heat-kernel expansion. *Class. Quantum Grav.*, **6**, 759–765.
Moss, I. G. & Dowker, J. S. (1989). The correct B_4 coefficient. *Phys. Lett.*, **B 229**, 261–263.
Moss, I. G. & Poletti, S. (1994). Conformal anomalies on Einstein spaces with boundary. *Phys. Lett.*, **B 333**, 326–330.
Moss, I. G. (1996). *Quantum Theory, Black Holes and Inflation.* New York: John Wiley & Sons.
Moss, I. G. & Silva, P. J. (1997). BRST invariant boundary conditions for gauge theories. *Phys. Rev.*, **D 55**, 1072–1078.
Mostafazadeh, A. (1994a). Supersymmetry and the Atiyah–Singer index theorem I: Peierls brackets, Green's functions and a proof of

the index theorem via Gaussian superdeterminants. *J. Math. Phys.*, **35**, 1095–1124.

Mostafazadeh, A. (1994b). Supersymmetry and the Atiyah–Singer index theorem II: the scalar curvature factor in the Schrödinger equation. *J. Math. Phys.*, **35**, 1125–1138.

Mostafazadeh, A. (1997). Parasupersymmetric quantum mechanics and indices of Fredholm operators. *Int. J. Mod. Phys.*, **A 12**, 2725–2739.

Musto, R., O'Raifeartaigh, L. & Wipf, A. (1986). The $U(1)$-anomaly, the non-compact index theorem, and the (supersymmetric) BA-effect. *Phys. Lett.*, **B 175**, 433–438.

Obukhov, Yu. N. (1982). Spectral geometry of the Riemann–Cartan space-time and the axial anomaly. *Phys. Lett.*, **B 108**, 308–310.

Obukhov, Yu. N. (1983). Spectral geometry of the Riemann–Cartan space-time. *Nucl. Phys.*, **B 212**, 237–254.

Olver, F. W. J. (1954). On Bessel functions of large order. *Phil. Trans. R. Soc. Lond.*, **A 247**, 328–368.

Paneitz, S. (1983). A quartic conformally covariant differential operator for arbitrary pseudo-Riemannian manifolds (preprint).

Parker, T. & Rosenberg, S. (1987). Invariants of conformal Laplacians. *J. Diff. Geom.*, **25**, 199–222.

Penrose, R. (1960). A spinor approach to general relativity. *Ann. Phys.*, **10**, 171–201.

Penrose, R. & Rindler, W. (1984). *Spinors and Space-Time, Vol. I: Two-Spinor Calculus and Relativistic Fields.* Cambridge: Cambridge University Press.

Penrose, R. & Rindler, W. (1986). *Spinors and Space-Time, Vol. II: Spinor and Twistor Methods in Space-Time Geometry.* Cambridge: Cambridge University Press.

Piazza, P. (1991). K-theory and index theory on manifolds with boundary, Ph.D. Thesis, M. I. T.

Piazza, P. (1993). On the index of elliptic operators on manifolds with boundary. *J. Funct. Anal.*, **117**, 308–359.

Polychronakos, A. P. (1987). Boundary conditions, vacuum quantum numbers and the index theorem. *Nucl. Phys.*, **B 283**, 268–294.

Reed, M. & Simon, B. (1975). *Methods of Modern Mathematical Physics, Vol. II: Fourier Analysis and Self-Adjointness.* New York: Academic.

Rennie, R. (1990). Geometry and topology of chiral anomalies in gauge theories. *Adv. Phys.*, **39**, 617–779.

Schleich, K. (1985). Semiclassical wave function of the universe at small three-geometries. *Phys. Rev.*, **D 32**, 1889–1898.

Schröder, M. (1989). On the Laplace operator with non-local boundary conditions and Bose condensation. *Rep. Math. Phys.*, **27**, 259–269.
Schwinger, J. (1951). On gauge invariance and vacuum polarization. *Phys. Rev.*, **82**, 664–679.
Seeley, R. T. (1967). Complex powers of an elliptic operator. *Amer. Math. Soc. Proc. Symp. Pure Math.*, **10**, 288–307.
Seiberg, N. & Witten, E. (1994). Monopoles, duality and chiral symmetry breaking in N=2 supersymmetric QCD. *Nucl. Phys.*, **B 431**, 484–550.
Shubin, M. A. (1987). *Pseudodifferential Operators and Spectral Theory.* Berlin: Springer-Verlag.
Sitenko, A. Yu. (1989). The $U(1)$ anomaly and zero-modes in two-dimensional Euclidean quantum electrodynamics. *Theor. Math. Phys.*, **81**, 1268–1279.
Spanier, E. H. (1966). *Algebraic Topology.* New York: McGraw Hill.
Sparnay, M. J. (1958). Measurements of attractive forces between flat plates. *Physica*, **24**, 751–764.
Stewart, J. M. (1991). *Advanced General Relativity.* Cambridge: Cambridge University Press.
Stewartson, K. & Waechter, R. T. (1971). On hearing the shape of a drum: further results. *Proc. Camb. Phil. Soc.*, **69**, 353–363.
Trautman, A. (1992). Spinors and the Dirac operator on hypersurfaces. I. General theory. *J. Math. Phys.*, **33**, 4011–4019.
Vafa, C. & Witten, E. (1984) Eigenvalue inequalities for fermions in gauge theories. *Commun. Math. Phys.*, **95**, 257–276.
Vancea, I. V. (1997). Observables of Euclidean supergravity. *Phys. Rev. Letters*, **79**, 3121–3124.
van den Berg, M. (1994). Heat content and Brownian motion for some regions with a fractal boundary. *Probab. Theory Related Fields*, **100**, 439–456.
van den Berg, M. & Gilkey, P. B. (1994). Heat content asymptotics of a Riemannian manifold with boundary. *J. Funct. Anal.*, **120**, 48–71.
van de Ven, A. E. M. (1997). Index-free heat-kernel coefficients (HEP-TH 9708152).
van Nieuwenhuizen, P. & Vermaseren, J. A. M. (1976). One-loop divergences in the quantum theory of supergravity. *Phys. Lett.*, **B 65**, 263–266.
van Nieuwenhuizen, P. (1981). Supergravity. *Phys. Rep.*, **68**, 189–398.
Varadhan, S. R. S. (1967). Diffusion processes in small time intervals. *Comm. Pure Appl. Math.*, **20**, 659–685.
Vassilevich, D. V. (1995a). Vector fields on a disk with mixed boundary conditions. *J. Math. Phys.*, **36**, 3174–3182.

Vassilevich, D. V. (1995b). QED on a curved background and on manifolds with boundary: unitarity versus covariance. *Phys. Rev.*, **D 52**, 999–1010.

Vassilevich, D. V. (1997). The Faddeev–Popov trick in the presence of boundaries (HEP-TH 9709182).

Waechter, R. T. (1972). On hearing the shape of a drum: an extension to higher dimensions. *Proc. Camb. Phil. Soc.*, **72**, 439–447.

Ward, R. S. & Wells, R. O. (1990). *Twistor Geometry and Field Theory.* Cambridge: Cambridge University Press.

Weidmann, J. (1980). *Linear Operators in Hilbert Spaces.* Berlin: Springer-Verlag.

Weyl, H. (1946). *The Classical Groups: Their Invariants and Representations.* Princeton: Princeton University Press.

Widom, H. (1980). A complete symbolic calculus for pseudodifferential operators. *Bull. Sci. Math.*, **104**, 19–63.

Wigner, E. P. (1959). *Group Theory and Its Application to the Quantum Mechanics of Atomic Spectra.* New York: Academic.

Will, C. M. (1993). *Theory and Experiment in Gravitational Physics.* Cambridge: Cambridge University Press.

Wipf, A. & Dürr, S. (1995) Gauge theories in a bag. *Nucl. Phys.*, **B 443**, 201–232.

Witten, E. (1982). Constraints on supersymmetry breaking. *Nucl. Phys.*, **B 202**, 253–316.

Witten, E. (1994). Monopoles and 4-manifolds. *Math. Res. Letters*, **1**, 769–796.

Wojchiechowski, K. P., Scott, S. G., Morchio, G. & Booss-Bavnbek, B. (1996). Grassmannian and boundary contribution to the ζ-determinant (TEKST NR 320).

Woodhouse, N. M. J. (1985). Real methods in twistor theory. *Class. Quantum Grav.*, **2**, 257–291.

Yajima, S. (1996). Evaluation of heat kernel in Riemann–Cartan space. *Class. Quantum Grav.*, **13**, 2423–2436.

Yajima, S. (1997). Evaluation of the heat kernel in Riemann–Cartan space using the covariant Taylor expansion method. *Class. Quantum Grav.*, **14**, 2853–2868.

York, J. W. (1986). Boundary terms in the action principles of general relativity. *Found. Phys.*, **16**, 249–257.

Index

Printed in the United States
By Bookmasters